The Pocket Darwin

Susan Brassfield Cogan

The Pocket Darwin
Copyright © 2016 by Susan Brassfield Cogan

All rights reserved.

No part of this book may be reproduced in any form or by any electronic or mechanical means including information storage and retrieval systems, without permission in writing from the author. the only exception is by a reviewer, who may quote short excerpts in a review.

ISBN-13: 978-1539343660
ISBN-10: 1539343669

Printed in the United States of America

Published by Coganbooks

On the Cover:
Portrait of Charles Darwin painted shortly after he returned from his voyage on the Beagle.

Cover and Book Design by:
Cogan Graphic Design
www.cogangraphicdesign.com

*For all their wonderful suggestions and corrections
I owe a very great debt of gratitude to*

Dr. Frank J. Sonleitner
Professor Emeritus
Department of Zoology, University of Oklahoma

Dr. Victor H. Hutchison
George Lynn Cross Research Professor Emeritus
Department of Zoology, University of Oklahoma

Dr. Rebecca Sherry
Interdisciplinary Perspectives on the Environment Program,
University of Oklahoma

*I am also grateful for the editorial assistance
of my friend, Debrah Hinton.*

ON

THE ORIGIN OF SPECIES

BY MEANS OF NATURAL SELECTION,

OR THE

PRESERVATION OF FAVOURED RACES IN THE STRUGGLE
FOR LIFE.

BY CHARLES DARWIN, M.A.,
FELLOW OF THE ROYAL, GEOLOGICAL, LINNÆAN, ETC., SOCIETIES;
AUTHOR OF 'JOURNAL OF RESEARCHES DURING H. M. S. BEAGLE'S VOYAGE
ROUND THE WORLD.'

LONDON:
JOHN MURRAY, ALBEMARLE STREET.
1859.

The right of Translation is reserved.

Table of Contents

1. Introduction	3
2. Charles Darwin: The Origin of an Idea	5
Darwin's Idea: The Struggle for Existence	
Alfred Russel Wallace	
3. What is Science?	10
What is a Theory?	
Scientific Prediction	
Proof, Truth and Certainty	
4. The Scientific Method	16
Peer Review	
5. The Scientific Method in Action	19
That Big Brain of Ours	
6. The Theory of Evolution	23
What is Evolution?	
Variation	
Mutations	
A Bug or A Feature?	
Natural Selection	
Sexual Selection	
Turning the Heads of the Females	
Kin Selection	
Genetic Drift	
Founder Effect	
Punctuated Equilibrium	
From Another Place	
Extinction	

7. Common Descent — 40
Taxonomy
DNA

8. Human Evolution — 47
The Missing Link: Human or Ape?
Australopithecus: The Taung Child and Lucy
Homo Erectus and Java Man
The Human Family Tree

9. The Fossil Record — 51
Transitional Fossils
Birds
Horses
Whales
The Bushy Tree
Living Fossils

10. The Age of the Earth — 59
How Do We Know the Earth is Old?
Radiocarbon Dating

11. The Beginning of Life — 62
What is life?
The Miller-Urey Experiment
The Primordial Organic Soup
Hydrothermal Vents

12. God and Darwin — 66

13. Bonus Essays
Two Opposable Thumbs Up for Evolution! — 71
"Evolution: It's Only a Theory"
And Other Famous Creationist Arguments — 83

14. References & Sources — 93

1. Introduction

Few people know what evolution is all about.

Some people think it is a progression to a higher and higher spiritual plane or a fight to become a "higher" animal and if you work really hard to improve yourself, you will become a more "evolved" human being. Some people think it is "survival of the fittest," so only the meanest, most selfish people or animals survive. Still others think it is just an excuse for atheists to ignore the Bible. They believe that evolution is not based on science, and nobody with any honesty or character would or should believe it.

Many people feel that life is vastly too complex to have evolved. Evolution is tiny. That's obvious. You can't take last year's flu shot because the flu virus evolves so quickly scientists have to develop a different vaccine every year to prevent infection. But the bird-flu virus isn't ever going to evolve into a chicken in a single step. Evolution is an accumulation of very tiny steps over a vast amount of time.

At this tiny level, almost everybody accepts evolution. A grain of sand is perfectly plausible. Everyone has seen a grain of sand. But can you believe that the Sahara Desert could be just grains of sand? It is so vast and complex it can't be explained as just billions of grains of sand. Right? Yet obviously, that is what it is.

Some people think evolution has something to do with the origin of life—it doesn't. Until something could make more of itself and unless it could do that with little variations there couldn't be evolution. Evolution couldn't start until after life began.

I hope to clear up all these confusing ideas about evolution with this little book. What is evolution? Is it scientific? What, exactly, did Charles Darwin give to the world? It would take more than a human lifetime to examine in detail all the evidence supporting the Theory of Evolution. This book will cover some of the main ideas of the theory and give pointers and references to find out more.

The Mudskipper: "Compared with fully aquatic gobies, these fish present a range of peculiar behavioural and physiological adaptations to an amphibious lifestyle. Anatomical and behavioural adaptations that allow them to move effectively on land as well as in the water. As their name implies, these fish use their fins to move around in a series of skips." From Wikipedia: https://en.wikipedia.org/wiki/Mudskipper (viewed September, 2016)

2. Charles Darwin: The Origin of an Idea

In September of 1837 Charles Darwin wrote "In July opened first notebook on transmutation of species. Had been greatly struck from about the previous March on character of South American fossils, and species on Galapagos Archipelago. These facts (especially latter), origin of all my views."[1]

Charles Darwin

When Darwin set sail on the Beagle in 1831, he was just 22 years old and had completed a degree in theology. He had every intention of returning in three years (instead of the five years the voyage actually took), to settle down to the life of a country pastor who collected beetles in his spare time. It didn't work out that way. His round-the-world voyage on the Beagle changed his entire life and the course of science history.

In that 1837 notebook, and in the many that followed, Darwin wrote down thousands of observations about the natural world. He noticed how species that lived near each

---

[1] *The Life and Letters of Charles Darwin* edited by Francis Darwin, 1905

other tended to be more similar to each other than species in the same genus that lived far away. He noticed a similar pattern in the fossils he collected.

Darwin's Idea: The Struggle for Existence

When Darwin originally collected the small black birds on the Galapagos islands he thought they were all different birds. They all had wildly different feeding habits. Some ate seeds, some ate insects and some pecked the backs of seabirds and drank their blood.[2]

> The Galapagos Islands are too far from the mainland of South America for birds to fly from one to the other. How did they make it over? One hypothesis is that they were blown there by hurricanes. Another hypothesis was that there once was another island halfway to the mainland. Geologists have confirmed this second hypothesis. The Galapagos Islands are volcanic. Lava wells up from the sea floor making an island which then gets carried away east by tectonic action. Eventually it sinks back beneath the waves. This has happened over and over creating a series of "stepping stones" between the mainland and the Galapagos. (Weiner, ibid, p.302-303)

The number of specimens of all kinds that Darwin collected during his voyage was huge, far too much for him to study in his lifetime. Therefore he gave parts of the collection to other scientists to study. John Gould got the little black birds. Gould discovered that the birds represented thirteen species of finches and they all were closely related to a single species on the mainland of South America, about 600 miles away. It looked very much like a few of the original South American species had made their way to the Galapagos.[3] After the ancestor birds arrived

---

[2] *Beak of the Finch: A Story of Evolution in Our Time* by Johnathan Weiner, 1995, p180
[3] Weiner, ibid.

on the islands they reproduced and their descendants diverged into the thirteen species known today. Darwin would have said they "transmuted"; we would say they evolved. If species change through

1. Geospiza magnirostris.
2. Geospiza fortis.
3. Geospiza parvula.
4. Certhidea olivases.

Large ground finch, Medium ground finch, Small tree finch, Green warbler-finch

time, the next question was: "How do they change?" A clue to this puzzle was given to Darwin by a book he read sometime in 1938. Tomas Malthus's *Essay on the Principle of Population* (1798) pointed out that many more organisms are born than can possibly survive and this causes a competition for resources. Malthus's ideas and conclusions were very different from Darwin's but Malthus's insights into the nature of reproduction and population led Darwin into thinking about what he called the struggle for existence.

> If there were any selective agency at work, it seems impossible to assign any limit to the complexity and beauty of the adaptive structures, which *might* thus be produced: for certainly the limit of possible variation of organic beings, either in a wild or domestic state, is not known. It was then shown, from the geometrically increasing tendency of each species to multiply (as evidenced from what we know of mankind and of other animals when favoured by circumstances), and from the means of subsistence of each species on an *average* remaining constant, that during some part of the life of each, or during every few generations, there must be a severe struggle for existence; and that less than a grain in the balance will determine which individuals shall live and which perish.—1844 Essay, Charles Darwin

Darwin spent twenty years researching and collecting evidence to support his theory of "transmutation of species." He hesitated to publish his material—he knew it was a hot topic—so he made plans for the material to be published in the event of his sudden death.

His books on the geology and paleontology of South America, coral reefs, and barnacles gave him a reputation as a first class scientist. In 1854 he began what he called his "big book," a gigantic multi-volume tome that would detail his theory and all the evidence supporting it. In May of 1858, Darwin remarked in a letter to Joseph Hooker: "There is not least hurry in world about my M.S."[4] In June of that year, that attitude changed dramatically.

Alfred Russel Wallace

On June 18, 1858 Darwin received an essay from a young naturalist named Alfred Russel Wallace. *On the Tendency of Varieties to Depart Indefinitely From the Original Type* set out most of the ideas about variation and natural selection that Darwin himself had developed fifteen years prior. Like Darwin, Wallace was also a self-trained naturalist. He also had traveled the world collecting specimens and making observations. He had even read Malthus. His greatest interest was in the geographical distribution of species and his specialty for most of his life was zoogeography. For two years he had been corresponding with Darwin on "the species question" and when the idea of vari-

Alfred Wallace

---

[4]*Darwin Correspondence Project*, Introduction to Volume 7, Frederick Burkhardt ed., Cambridge UP, 1991

ation and natural selection came to him he naturally sent his essay on the subject to Darwin asking for help to get it published.

This Darwin did, but not in the way Wallace expected. When Darwin read Wallace's essay, he realized Wallace had independently come up with the same ideas Darwin had written in his 1944 essay—fourteen years earlier! Even so, Darwin knew he had to share credit for the idea with Wallace. In July of that same year, Wallace's essay and some of Darwin's unpublished writings on the same subject were read as "The Darwin-Wallace Paper" at the Linnean Society. [5]

The idea of evolution itself was not original with Charles Darwin, but the mechanism for evolution—variation and natural selection—was. His idea of being "not [in the] least hurry" was exploded. In order to get credit for his ideas he needed to get them into print and fast. He immediately began to make an abstract of his voluminous writings and research notes. This "abstract," published about eighteen months later, ran to 490 pages. It is now known as *On the Origin of Species by Means of Natural Selection, or the Preservation of Favoured Races in the Struggle for Life*. Its first printing sold out in a single day.

> "Races" in the title of *Origin* refers to what we would call today "subspecies." There is no discussion of human races or human evolution in Darwin's first book. He was not to write on that topic for another 20 years.

---

[5] The Darwin-Wallace Paper www.linnean.org (viewed September 9, 2016)

3. What is Science?

Perry Mason regards the standard-issue corpse in front of him. Della Street asks him a question.

"Well, Perry, do you have a theory as to how this happened?"

"Yes, Della, I do," he says.

No, he doesn't. In popular usage, a theory is a guess, a speculation. Not so in science. If Perry Mason has a scientist he would say "No, Della, I have a hypothesis." He has a possible explanation that will need to be supported—or not—by further investigation. Paul Drake will sleuth around, Perry will interview witnesses and family members. The bumbling Lieutenant Tragg might even make a contribution. Above all, facts will be gathered which either support Perry's hypothesis or not. If not, then he will be forced to make a new hypothesis to fit all the facts without ignoring any.

In the Perry Mason books, films and tv shows the foundational hypothesis—that the client is innocent—is always true. Science and nature are not so conveniently neat and tidy. If your hypothesis is always true no matter what the facts are, then you don't actually have a hypothesis. You have a belief. All scientists know they might be forced to face the fact some day that their client isn't innocent. That unhappy possibility is there every time a new fact is uncovered.

A hypothesis is a tentative idea that can be tested and possibly be proved false. The idea that a hypothesis must be able to be proved false is *very, very important*. Keep that in mind. That doesn't mean it *will* be proved false or has been. It just means it *could* be. You could have a hypothesis that gravity is caused by invisible pink unicorns holding us down on the ground, but how would you test it? How would you prove it true or false? Even if it were 100% true, science can't collect data for or against it, so the invisible pink unicorns will have to do their job without any credit.

What is a Theory?

Scientists sometimes use the word "theory" in an informal setting when they actually mean "hypothesis" just like Perry Mason does. It can get confusing. Basically a theory is a collection of hypotheses that have been supported by the evidence. When many hypotheses surrounding a general subject get validated, then your view of that subject will get very broad and deep. You will get a systematic model of how some particular part of nature works. The Theory of Evolution, the Theory of Gravity and the Germ Theory of Disease have many facts and many answered hypotheses supporting them.

Theories have practical use. Because they have been supported by so many facts and observations, theories can generally be relied on to provide answers to real-life problems.

For example, when HIV began to spread in the US, several people, mostly young gay men but also a few women and some children, came down with a rare cancer usually found only in old men called Kaposi's sarcoma. Also, some people contracted what had previously been a very rare form of pneumonia. There was no pattern and it was very confusing. Then several young men in Orange County, California became sick with Kaposi's sarcoma. As investigation went forward, it was discovered that all the young men were homosexual and many of them knew each other. Cancer is not normally communicable, but the way the rare diseases were spreading looked very much like a pattern that the germ theory of disease would predict. Researchers began to suspect early on that some kind of communicable disease was causing the rare cancers and pneumonias and within only a few years, the HIV virus was isolated.[6]

Once HIV was identified as a virus, Evolution Theory stepped in. HIV reproduces very rapidly and it has a very high mutation rate. That means it evolves very quickly. Anti-viral drugs have improved tremendously since the early days, but the first few developed would only work for a little while. HIV would rapidly evolve around anything that killed them directly. Most of the research on HIV has been a search for something HIV can't evade by evolving.[7]

Scientific Prediction

A scientific prediction is not the same kind of prediction you will get from Madame Zolar who sits gazing into a crystal ball with a little incense burning in the background.

Even though they have never been actually observed by humans, it is a fact that atoms exist. We know this because atomic theory predicts that if atoms exist they will behave in certain ways. That is a scientific prediction.

In 1862, someone sent Darwin an orchid from Madagascar which had a nectar tube that was a foot deep. Darwin predicted that some kind of insect would be found which would be able to pollinate such an odd flower. Forty-one years later, after Darwin had been dead for more than twenty years, a moth was discovered with a long, thin probos-

---

[6] *The AIDS Crisis, A Documentary History* by Douglas A. Feldman, Julia Wang Miller, 1998
[7] HIV Evolution and Escape *Trans Am Clin Climatol Assoc.* 2004; 115: 289–303.

cis which made its living sucking the nectar of the orchid, spreading the orchid's pollen from flower to flower.

Eventually enough hypotheses about a certain subject will pan out or be rewritten until they fit all the known facts and these hypotheses will be elevated to a theory. The Theory of Evolution is actually a collection of very successful hypotheses.

Like a hypothesis, a theory could possibly be proved false—and occasionally they are—but generally if a system of ideas is called a theory it has so much going for it, it is unlikely that it is wrong.

Atomic Theory predicts that certain kinds of atoms will be radioactive. Nuclear power plants are designed on the basis of Atomic Theory; let us hope it is not "only" a theory!

Evolution is a fact. The Theory of Evolution is a system of ideas explaining all the many observations science has made about evolution. The Theory of Evolution has had a powerful impact on all the biological sciences, including agriculture and medicine, and yet it is still "only" a theory.

There are several things which, if true, would prove evolution to be false. The most famous one was proposed by J. B. S. Haldane. He suggested that if they found rabbit bones among Cambrian era fossils it would be evidence against evolution. The Cambrian era was long before the Jurassic era of the dinosaurs and that was long before the Cretaceous era when mammals like rabbits began to appear.

When Darwin first wrote about evolution, many scientists already believed species change over time. What Darwin proposed was an explanation for *how* that could happen. In the beginning, while he was still collecting data to support his theory, his ideas really only amounted to a hypothesis, a hypothesis that could have been proved false if the data had not been available to support it. Over the last century and a half, some of the details of Darwin's theory have been refined and augmented with new information, but generally, Darwin was correct. So far the data have always supported the Theory of Evolution, even data that were discovered long after Darwin lived.

Proof, Truth and Certainty

Science never deals in proof, as such. It deals in evidence. All scientific conclusions must stand ready to be rewritten when new evidence is discovered. When a news source proclaims "science as proved (or disproved) XYZ!" you can be sure the scientists involved did not phrase their conclusions that way. They said "this study suggests XYZ may (or may not) be the case." The scientists know full well that a future study may suggest something different. It is the news media that don't understand that.

Theories are not proved. They are supported by the preponderance of the evidence—or not. Though "proof" is a word scientists will sometimes use when speaking informally, it is not something that scientists talk about when they are speaking formally of the evidence supporting any scientific theory. This provisional attitude allows scientists to refine and expand our knowledge. Science always assumes that there is more to know and more to understand, that we can never know and understand it all. Therefore, all of scientific knowledge is held provisionally and can be discarded if new data shows that old ideas and theories are wrong. Who would want it any other way? The truth of the matter is really all that's important.

> "Well evolution is a theory. It is also a fact. And facts and theories are different things, not rungs in a hierarchy of increasing certainty. Facts are the world's data. Theories are structures of ideas that explain and interpret facts. Facts do not go away when scientists debate rival theories to explain them. Einstein's theory of grav-

itation replaced Newton's, but apples did not suspend themselves in mid-air, pending the outcome. And humans evolved from apelike ancestors whether they did so by Darwin's proposed mechanism or by some other yet to be discovered."[8]

Well-supported theories are generally treated as part of human knowledge. Details of theories will change from time to time as new observations are made, but generally a theory is considered foundational. This is true of Atomic Theory, the Germ Theory of Disease, the Theory of Relativity and the Theory of Evolution.

Above all else, though, science isn't an entity in itself. It's not a person or collection of people. It's not a military-industrial complex. Science is a method, a method for collecting and making sense of the "world's data." It's a method anybody can use and it can be used in almost any situation. It may not help you arrive at a final Truth™ but it will help you get closer.

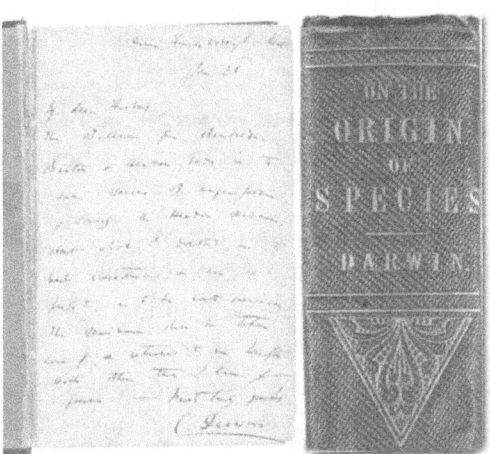

---

[8]"Evolution as Fact and Theory," *Hen's Teeth and Horse's Toes*, Stephen J. Gould, New York: W. W. Norton, 1994, p. 254.

4. The Scientific Method

People wonder how scientists can know what they know. How did we learn so much about life's history? Nobody was there to watch it happen. How do we know electrons exist? Nobody has ever seen an electron. How do we know the Big Bang happened? Nobody could possibly have witnessed the birth of the universe.

It's impossible to understand evolution or any branch of science without understanding the scientific method.[9] The scientific method is the best way developed thus far for discovering and exploring practical truth usable in the everyday world.

Scientists use this simple but powerful method to discover how the universe, the world, and our own bodies work. This is a general overview:
1. Make some observations about the world.
2. Make a hypothesis, a tentative explanation, that is consistent with what you have observed and which can, at least in theory, be proved wrong.
3. Use the hypothesis to make predictions about facts that might be observed if the hypothesis is true.

---

[9] http://www.sciencebuddies.org/science-fair-projects/project_scientific_method.shtml (viewed September, 2016)

4. Test those predictions by experiments or further observations.
5. If necessary, adjust your hypothesis to fit the new observations you have made.
6. Go back to step 3 and repeat until your hypothesis matches all the known facts and the results of your experiments and observations and isn't contradicted by any facts that you know of.

As more and more evidence accumulates to support a hypothesis and as it is refined to fit with observations, gradually the hypothesis can begin to be fleshed out into a theory.[10] However, the relationship between hypothesis, theory, and "law" is not as linear or quite as simple as can be presented here.

Peer Review

The above outline of the scientific method, leaves out peer review. Peer review is an important part of the scientific method, because it is a way to detect and correct mistakes. When a scientist is beginning to be confident that her hypothesis is true, she will submit a paper for publication describing it, all the relevant observations that support it, and the methods used to make those observations. Other scientists review the paper and decide if the methods used to collect the data were valid and if the conclusions made in the paper are warranted by the evidence. If the paper meets that standard, then (sometimes) it is published.

It is impossible for every scientist to do every experiment independently to confirm every hypothesis. Because life is short, scientists must trust each other. A scientist who claims

---

[10] http://teacher.nsrl.rochester.edu/phy_labs/AppendixE/AppendixE.html (viewed September 2016)

to have done an experiment and obtained certain results will usually be believed, and few people will bother to repeat the entire experiment. Parts of experiments get repeated as part of other experiments, though. Most scientific papers contain suggestions for other scientists to follow up. Usually the first step in following up is to repeat a part of the earlier work. So if a hypothesis is the starting point for a significant amount of work then the initial experiments will get repeated many times.

Peer review doesn't always work. Peer review examines methods and conclusions, but the basic assumption is that the scientist is reporting those things honestly. Scientists are human beings and, unfortunately, not always honest. Thankfully, the scientific method has a way of exposing fraud just as it exposes honest mistakes.

5. The Scientific Method in Action

In the early days of paleoanthropology, naturalists had a working hypothesis that the big brains of humans evolved first and upright walking, tool use, and everything else that makes us human evolved afterward. This was not an unreasonable idea considering that in Darwin's time the only hominid fossils were some specimens of Neanderthal who had big brains but a lot of primitive features.

Charles Dawson (PD-US)*

The discovery of some fossils at a rock quarry south of London near Piltdown, beautifully and perfectly seemed to support the big-brain-first hypothesis. The fossils, discovered in 1908 and 1913 by Charles Dawson, had a large head and a very primitive jaw. It was beautiful. It was perfect. It was a fraud.

So far the scientists had done everything right. They had made a hypothesis—the big brains of humans evolved first—and they had made an observation, which seemed to support that hypothesis—the fossils, which were named Piltdown Man after the area where they were found. They also fit a prediction arising from that hypothesis, that Hominid fossils older than Neanderthal would have big brains. At

*https://en.wikipedia.org/w/index.php?curid=13322116

that time there was no way to directly date the Piltdown fossils, but they were found with some genuine fossils from the Pleistocene, which had been planted to add credibility. To make matters worse, all the things that would have helped verify the fossils' authenticity were missing. The joints that would have shown how the jaw fit to the skull had been broken off. The teeth had been filed to make them look worn down so the markers that would have distinguished them as orangutan were missing. They had even been stained with fluorine to make them look old.

None of that would fool modern geologists and paleontologists, but those sciences were in their infancy when Piltdown Man was found. Also a very human reason led to the acceptance of Piltdown. France had lots of fossil hominids. England had none. Even national pride, by itself, would not have been enough motive to accept the fossils if they had not confirmed the prevailing hypothesis that humans developed big brains first.

Piltdown man was greeted with great joy and much publicity. There were quite a few doubters at first, but when the second set of fossils was discovered in 1913, most of the doubters were convinced. Casts were made of the fossils for study, and the original bones were locked in a safe for more than 40 years.

All this sounds like the scientific method didn't work. The scientists had an idea, they found a piece of evidence that confirmed their idea, but the evidence was *false*. How did the scientific method save the situation? Steps #4 (find evidence which fits your predictions) and #5 (change your hypothesis, if necessary, to fit the evidence) are the toughest of all. It's difficult to let go of an idea, even when the evidence is against it, but the significance of Piltdown dwindled as new evidence came to light.[11]

---

[11] https://en.wikipedia.org/wiki/Piltdown_Man (viewed September 2016)

That Big Brain of Ours

At the time Piltdown Man was discovered there were not many hominid specimens to compare it to. A few Neanderthals had been found and they had very large brains. This was a point in favor of the big-brains-first hypothesis. In 1891 Eugene Dubois found an example of *Homo Erectus* which had a small brain. It was only a skullcap and not well received by the scientific community. In the 1930s others found more *Erectus* fossils, all with very small brains and fairly modern bodies. All of them were clearly upright walkers. *Australopithecus africanus*, found in 1925, lived much further back in time than *H. erectus*. It had human-like teeth and the position of the hole at the base of the skull indicated that it walked upright, but it, too, had a very small brain.

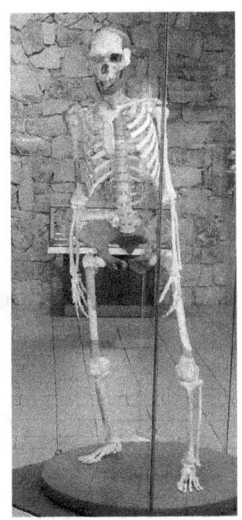

Homo erectus from Tautavel, France*

As the years went by, more and more fossils were found, not only many more examples of *Homo erectus* but also an *Australopithecus africanus* in 1936 and an *Australopithecus robustus* in 1938. It was becoming clear that the "big-brain-first" hypothesis was wrong. Early human ancestors walked upright first and the brain developed slowly over several million years.

These discoveries (observations) left Piltdown Man more and more out in the cold. By the late 1940s nobody thought Piltdown was a fraud—scientists trust each other—but it was clearly some kind of mistake. No other observation confirmed it.

*By Gerbil - Own work, CC BY-SA 3.0, https://commons.wikimedia.org/w/index.php?curid=4800669

Now the scientific method was turned to Piltdown Man itself. What did those fossils represent? The working hypothesis was that the fossils had been misdated. They had to be fairly recent. In the late 1940s a fluorine test was finally available and the fossils were tested for age. They were relatively young. It was also determined that geologically that the Piltdown gravel pit could not be from the Pleistocene. Okay, scientists thought, they were definitely a mistake of some kind.

In July of 1953, J. S. Weiner, an eminent paleontologist, saw the Piltdown fossils in a context with several other hominid fossils and it occurred to him that the Piltdown fossils might be a hoax. Once the possibility was raised, showing it to be true was simple. No one had ever thought to look at the teeth under a microscope—after all, no one suspected outright fraud. When Weiner did that, he saw the signs of artificial abrasion. The teeth had been filed to make them look worn.

Piltdown Man had contradicted too much of the other evidence surrounding it, so the hypothesis it represented—that it was a representative of an early hominid with a big brain—had to be rewritten.

I suppose the $64,000 question at this point is who perpetrated this fraud upon the scientific community? Unfortunately there is only a $.05 answer. Nobody knows for sure. The chief suspect, of course, is Charles Dawson who "discovered" the original fossils. The problem with him as a suspect is, though he wasn't a trained scientist, he had an excellent reputation in the scientific community. Also he died in 1914, a year after the second Piltdown discoveries. Several other names have been proposed and everyone had or has their pet theories. But nobody knows for sure, and it's unlikely anybody will ever know.

6. The Theory of Evolution

What is Evolution?
Darwin only used the word "evolved" once in the 6th edition of *The Origin of Species*.

> There is grandeur in this view of life, with its several powers, having been originally breathed into a few forms or into one; and that, whilst this planet has gone cycling on according to the fixed law of gravity, from so simple a beginning endless forms most beautiful and most wonderful have been, and are being, evolved.

Darwin called his theory "descent with modification." Modern population geneticists define evolution as a change in the gene pool of a population over time. It is a simple, yet powerful idea. The Theory of Evolution seeks to explain how gene pools change and why they change.

> *"It is populations that evolve, not individuals."*

A gene is a unit of heredity that passes from generation to generation. The gene pool is the set of all genes in a species or a population. It is populations that evolve, not individuals.

In the 1980s scientists discovered that one particular Italian village had far fewer heart attacks and strokes than the general population. Many of the people of the village had a mutated protein that prevented cholesterol build up in the arteries. The gene for that protein was traced back to a single man who lived more than 200 years ago.

Did that man "evolve"? No. An individual can't evolve. He had a lucky mutation, which he could then pass to his descendants. As the mutation spread through the village, the population as a whole evolved over time.

Darwin did not know about genes. He could see that offspring inherited physical traits from their parents, but he did not know how. During Darwin's lifetime, Gregor Mendel was doing important research into inheritance, but Darwin never knew about it. Mendel's paper on inheritance was published in an obscure journal called *The Proceedings of the Natural History Society of Brunn* where it was ignored for 35 years. Nobody understood its importance. It is rumored that after Darwin died, Mendel's paper was discovered in Darwin's study, still unopened, but that has not been substantiated. In 1900, 18 years after Darwin's death, Carl Correns, Hugo de Vries, and Erich von Tschermak rediscovered Mendel's work and the mechanism for inheritance that had always eluded Darwin.

Gregor Mendel

Variation

When you have a litter of puppies in front of you, one of the first things you notice is how different they are from each other in size, color and personality. Variation was the first thing Darwin noticed when he began observing plants and animals in domestication and in the wild.

Genetic variation is the raw material of evolution. Some puppies in any given litter are larger, some smaller, some fatter, some thinner, some lighter, some darker. Most of these variations are neutral and will have no effect on the puppies' survival. However, if external conditions change, if there is a food shortage, for example, the fatter ones will survive better if the fatness means they metabolize food more efficiently. If the puppies find themselves in snowy conditions, lighter fur might help them hide from predators or allow them to sneak up on prey. Whether or not a variation is an advantage depends, for the most part, on the environment.

Mutations

When a cell divides, the genes duplicate themselves so that each cell will have an identical copy of all the information it needs to perform whatever function it has in the body. That duplication is almost always perfect. But, sometimes, for many reasons, including some that are unknown, the gene will not be copied perfectly. This non-exact copy is called a mutation. It is mutations that give rise to the variation necessary for evolution. Many mutations are neutral and will not harm the organism at all. Some mutations are harmful, and death or impaired function results. Some mutations, though, are lucky. They help the individual and its descendants exploit a new food source, escape a predator, resist disease etc. They will give an individual an advantage in the "struggle for life" and that advantage will be passed on to their offspring. In fact, a lucky advantage will allow them to leave more offspring than other members of their population. Eventually that lucky mutation and its advantage—whatever it is—will spread throughout the population and that population will have evolved.

A Bug or A Feature?

Mutations are sometimes characterized as a bug in the system, they are "mistakes" or "accidents"—but are they? Yes, the genes that produced them did not duplicate themselves exactly, but a population without variation isn't going to last long.

Let's imagine a population whose genes replicate without any "mistakes." Let's call them Furry Critters. All Furry Critters look alike because their genes replicate perfectly at all times. The Furry Critter resembles a meercat and lives in a grassland which supplies all its needs.

The entire world isn't covered in a congenial grassland, though. If Furry Critter lives near a woodland, most of the potential food sources in the woodland would be closed to it. If it lived near water, it would not be able to use that food source very well either. It would have difficulty surviving near a desert or a glacier. All that doesn't matter as long as Furry Critter is perfectly suited to its habitat, the grassland.

Then things change. What if the Sahara advances south? It's been doing that for a few centuries now, turning former grasslands into dune desert. Or maybe an ice age happens. There have been at least four ice ages in the history of the earth. One of the things you can count on in this world is that things don't stay the same. What happens to our perfect, non-varying population if the Sahara decides to expand to the south and they are in its way? They go extinct. If there is no variation in their genome, if their genes have been replicating perfectly without a "mistake," then there are no individual with any characteristics that might help them survive in a desert. There is no variation for natural selection to select, so it selects none of them and they go extinct.

Clearly, since the world changes all the time, the tendency of genes to copy themselves imperfectly is a big advantage. It's a feature of the system, not a bug. If there is variation, and some of our Furry Critters were tolerant to heat and drought, they will survive better than their companions who must have cool weather and lots of water. The unlucky Furry Critters who were perfect for their environment will face disaster if their environment changes. Over time the unlucky members of the group will die or leave fewer offspring and the more desert-hardy ones will live and leave more offspring. Eventually all the new Furry Critters will be perfectly suited to the dryer climate. They will have evolved.

Natural Selection

The main mechanism Darwin hypothesized for evolution was variation plus natural selection. He did a thorough study of human selection that goes on in plant and animal breeding during the time that he was gathering evidence for his theory. He corresponded with men who developed new breeds of cattle, sheep and dogs. He even began to raise pigeons himself. He saw how you could get rapid changes in an animal's body by simply removing any individual from the population that did not have the desired trait.

Darwin hypothesized that the environment "selects" organisms the same way. It selects organisms that have everything they need to survive in their current habitat, or at least have enough advantages to pass their genes on to the next generation.

> [W]e may feel sure that any variation in the least degree injurious would be rigidly destroyed. This preservation of favourable

individual differences and variations, and the destruction of those which are injurious, I have called Natural Selection.[12]

Less fit organisms die off or leave few, if any, offspring. The more fit an organism is, the more offspring it leaves behind. What "fit" means changes as the environment changes. Furry Critter was perfectly suited to its environment. Too perfectly suited. Nevertheless it was fittest as long as the environment was stable.

> Humans are a part of nature and are, along with some insects, a predator species from the point of view of broccoli, cauliflower, etc. Therefore the selection is natural. However, most people think of it as "artificial" selection since it's humans doing it. In ancient times, however, it was not known that simply choosing what you like out of a patch of plants and plowing the rest under would cause evolution.

It is easy to see how much change can be brought about when humans are doing the selecting. Broccoli, cauliflower, cabbage, kale, and Brussels sprouts all came from one species of a wild mustard through the efforts of human selection. This is an example of extremely rapid evolution. Natural selection sometimes works this quickly, but is usually much slower and more gradual.

> It may be said that natural selection is daily and hourly scrutinising, throughout the world, every variation, even the slightest; rejecting that which is bad, preserving and adding up all that is good; silently and insensibly working, whenever and wherever opportunity offers, at the improvement of each organic being in relation to its organic and inorganic conditions of life. We see nothing of these slow changes in progress, until the hand of time has marked the long lapses of ages, and then so imperfect is our view into long

---

[12] *Origin of Species*, Charles Darwin, Chapter 4

past geological ages, that we only see that the forms of life are now different from what they formerly were.[13]

Sexual Selection

Males of many species develop very distinct secondary sexual characteristics. Some examples are the peacock's tail, the horns of elk and cattle, bird calls, and flashes in fireflies.

> Sexual selection depends on the success of certain individuals over others of the same sex, in relation to the propagation of the species; while natural selection depends on the success of both sexes, at all ages, in relation to the general conditions of life.[14]

Many of these traits are actually bad from the standpoint of survival for the individual organism, because in come cases it takes a great deal of additional energy to produce and use these babe magnets. Also, the strategy which catches the eyes of the ladies may also attract the attention of a predator. Nevertheless the advantages to the male's genes outweighs the disadvantages. A male who lives a short time, but produces many offspring is much more successful (in an evolutionary sense) than a male who lives a long time, but leaves few offspring. The genes of successful males will eventually spread throughout the gene pool of that particular population.

Turning the heads of the females

A fun example of sexual selection is the Bowerbird. Several kinds of bowerbirds live in Australia and New Guinea. For most of the year males and females live apart, but during breeding season the males gather to build their bowers and attract mates.

---

[13] *Origin of Species*, Chapter 4
[14] *Descent of Man*, Charles Darwin, 1871.

The bowerbird builds various kinds of structures out of leaves and twigs, which it decorates with all kinds of shiny, pretty objects like stones, feathers, colorful scraps of paper or bits of aluminum foil discarded by humans. Sometimes the decorations are stolen from the bowers of the competition! Some species of bowerbirds build a little U-shaped corral while others build platforms ringed with pretty leaves, which are replaced with fresh ones as they wilt. A species called Macgregor's Bowerbird (*Amblyornis macgregoriae*) builds a sort of decorated "maypole" which is sometimes painted with berry juice mixed with their own saliva. In addition to these structures male bowerbirds also sing, dance and pose for their prospective mates.

The females do the choosing and usually the older, more experienced male birds are the winners in the mating game.

The cost of this behavior is high. All the time the birds are building their avian equivalent of bachelor pads, they must hunt for food and avoid predators. It's tough, but rewarding. A successful male bowerbird will mate with dozens of females in a single breeding season and therefore father a great many offspring, which carry his successful genes.[15]

Kin Selection

John Maynard Smith coined the term "kin selection" in the 1930s. It is an interesting feature of evolution that not a lot of people know about. Think about this: Your brother is trapped in a burning building. Knowing that you might be killed by the fire, you rush in to save him. What's up with that? Evolution, so people say, is about survival of the fittest. It's about the bigger, stronger individuals surviving and the weaker, less lucky members of the group dying or leaving fewer offspring. Why would someone rush into a building

---

[15] https://en.wikipedia.org/wiki/Bowerbird (accessed September 13, 2016)

to save their brother? Or their cousin? Or a total stranger? Kin selection is the answer.

Bees will pass on information about a food source to other bees without taking any of the food themselves. A prairie dog will chirp to alert its tribe of a predator, possibly attracting the attention of the predator to itself, but saving the whole group. A quail hen will invite a predator to chase her in order to lead it away from her chicks.

These examples of self-sacrifice are extremely common. They are easy to find. When Darwin first began to study the problem of altruism in nature he was puzzled. Surely natural selection would favor individuals who looked out for their own survival and not those of their kin.

He studied the behavior of ants and ant drones (or "slaves" as he called them). In *Origin of Species*, he noted that the drones were sterile and could not pass on their genes but nevertheless were vital to the survival of the hive, all of whom were their kin.

In *The Descent of Man* he also touched on kin selection briefly in his discussion of the evolution of morality:

> It must not be forgotten that although a high standard of morality gives but a slight or no advantage to each individual man and his children over the other men of the same tribe, yet that an increase in the number of well-endowed men and an advancement in the standard of morality will certainly give an immense advantage to one tribe over another. A tribe including many members who, from possessing in a high degree the spirit of patriotism, fidelity, obedience, courage, and sympathy, were always ready to aid one another, and to sacrifice themselves for the common good, would be victorious over most other tribes; and this would be natural selection. [16]

---

[16] *The Descent of Man, and Selection in Relation to Sex,* Charles Darwin, 1871

Since Darwin's time, altruism and kin selection have been studied intensively. W.D Hamilton[17] was the foremost authority on kin selection. He showed, through a mathematical study of genetics, how kin selection works.

You share about 1/4 of your genes with your cousin. Nobody rushes into a burning building shouting "Jim shares one quarter of his genes with me. I must save him!" We have a "natural" (and therefore at least partially instinctive) tendency to look after our kin. Humans didn't know anything about genes while we were evolving. That's a new thing that's only about a hundred years old. The impulse to take care of each other evolved without that knowledge because it carries a strong evolutionary advantage.[18]

> A man has an X and a Y chromosome. A woman has two X chromosomes and no Y. Mom can only contribute an X chromosome to her children. Dad can contribute either an X or a Y. If Dad contributes an X chromosome to the baby he will have a daughter. If he contributes a Y, he will have a son. Statistically, X's and Y's get passed on at about the same rate. If a father has four daughters and no sons, none of the genes on his Y chromosome will get passed to his descendants.

Genetic Drift

Since Darwin did not know about genes, he could not know about genetic drift. Genetic drift is a modern refinement to his theory. In fact, at first glance genetic drift might seem to contradict Darwin since it is about evolution that occurs without natural selection.

Every generation is unique. Parents don't always pass on all their genes to their offspring. Just through general statistical sampling some genes get passed on to the next

---
[17]https://en.wikipedia.org/wiki/W._D._Hamilton (accessed September 13, 2016)
[18]The Selfish Gene, Richard Dawkins, 1990, p. 93

generation and some genes don't. In a large population, that doesn't matter much because, since most individuals have most genes, the gain or loss will average out.

Genetic drift becomes really important when the population gets small or when a gene is rare. The odds of some individual carrying a certain gene go down. Natural selection doesn't have much to do with it.

Let's go back to our population of Furry Critters. Let's say some Furry Critters are gold, some are silver. A few are mottled silver and gold. One day they are all sunning themselves on the rocks and a big avalanche rains down killing quite a lot of them. After the disaster, when they sort themselves out they notice that purely by accident, almost all the golds and most of the mottled ones were killed. Only a few of the silver ones were killed.

The avalanche was not natural selection, as such. There was no particular advantage to being silver. The silver-colored Furry Critters just happened to hit it lucky. There are now far fewer gold genes than silver genes in the population.

Some Furry Critters still have the gold gene but it is now more rare than it was and pure random sampling may eventually cause it to be lost to the Furry Critter population. If being gold also gave some kind of advantage—let's say it made it easier to hide from predators—that advantage is also lost. [19]

Founder Effect

Another thing that happens to a small population is what scientists call the Founder Effect.

Let's say a thousand years after the avalanche the Furry Critters experience an earthquake and a big crack opens up in the earth. Most of the Furry Critters are on the west side of the crack and a few are on the east side. By now all Furry Crit-

---
[19] *Evolution*, Monroe W. Strickberger, 1990, p. 515

ters are silver and the gold gene is extremely rare. However, by chance, the east-side Furry Critters happen to have more of the gold gene than the general population. The gold gene now has a greater chance to be expressed in the population. If it did give some kind of advantage it will be much more easy for it to spread in a small population than a large one.

Sometimes the Founder Effect doesn't leave enough variation and the population has fewer options to offer natural selection.

This is what happened during the Irish potato famine in the early 19th century. All Irish potatoes were descended from only four types that had been imported from the Americas. This meant they had so little genetic diversity that a single fungus—called the blight—could destroy all of them. If there had been a broader genetic base (and remember, in 1840 nobody knew about genes) then perhaps some potato plants would have been resistant to the blight.

Four kinds of potato more or less randomly ended up in Ireland. This caused the low diversity that is the Founder Effect. The Founder Effect in this case, working in combination with natural selection, proved to be disastrous.[20]

Another example of the Founder Effect is the cheetah. About 10,000 years ago something caused a genetic bottleneck in the cheetah population. Somehow, all but a few individuals were wiped out. As it happens, a lot of large vertebrates went extinct at the same time but nobody knows why. All modern cheetahs descended from the few who survived. There are now two subspecies of cheetah but there is very little genetic diversity among them. Fortunately for them, their environment is very stable and, like Furry Critter, they are well suited to their environment.

---

[20]https://en.wikipedia.org/wiki/Great_Famine_(Ireland) (accessed September 13, 2016)

Human beings also experienced a genetic bottleneck about 70,000 years ago. All humans are descended from perhaps 10,000 individuals and some say even fewer. We are 99.98% genetically identical. We don't even have any sub-species. All human differences are very superficial and hardly more than skin deep.

How do scientists know all this? They study the DNA of the populations involved. Neutral mutations occur at a fairly steady rate and so geneticists can compare the genomes of two populations, count the mutations and determine how long the populations have been separate. They can't say for sure that a genetic bottleneck occurred on November 3, 68,483 BC, but they can give a rough idea of how long ago the most recent common ancestor lived.[21]

Punctuated Equilibrium

Sometimes evolution goes fast. Sometimes evolution goes slow. When you look at the way fossils are found in the geological layers, you occasionally find extremely gradual change. But fairly often, new fossils appear suddenly in the geological record.

When a geologist says the word "sudden" though, you shouldn't be thinking of a stage magician causing a pigeon to materialize out of a silk scarf. When a geologist says "sudden" they mean maybe a 10 thousand years. Ten thousand years is a very, very long time compared to the short life of a human being. It's such a long time, it's almost difficult for us to imagine it. But to a geologist even a million years is an eye-blink.

Punctuated equilibrium was an idea developed by Stephen J. Gould and Niles Eldridge in 1972[22] to explain the

---

[21] Strickberger, ibid., p. 517-518
[22] Eldredge, N., & Gould, S. J. (1972). Punctuated equilibria: an alternative to phyletic. In: *Models In Paleobiology*

relatively sudden appearance of new species in the fossil record. Since fossilization is rare it gives us a very imperfect sample of all the species that ever lived. At one time there were millions of passenger pigeons in the United States. Now there are none. As far as anyone knows, they are extinct. Yet there are no fossils to indicate to future generations that they ever existed.

A large, stable population is more likely to leave a few fossils behind than a population which is small and rapidly evolving. So if you have a new, small population of Furry Critters the odds are pretty good they will have to grow into a big population before they leave even a single fossil. From our point of view, that makes them seem to pop into existence out of nowhere.

Gould and Eldridge pointed out that the fossil record does not always show gradual change from one species to another. They accused Darwin of believing in such gradual fine-grained change. It was a bad rap. Darwin believed no such thing.

> Only a small portion of the world has been geologically explored. Only organic beings of certain classes can be preserved in a fossil condition, at least in any great number. Widely ranging species vary most, and varieties are often at first local, — both causes rendering the discovery of intermediate links less likely. Local varieties will not spread into other and distant regions until they are considerably modified and improved; and when they do spread, if discovered in a geological formation, they will appear as if suddenly created there, and will be simply classed as new species.[23]

Even Darwin understood that evolution could sometimes be, or at least seem, very rapid.

---

[23] *Origin of Species*, Charles Darwin, 1st Edition 1859, p.439

From Another Place

Eldredge and Gould's explanation is based the idea that a small portion of a population will get isolated from the main group causing allopatric speciation. Allopatric means "from another place." So if a group becomes isolated from the parent species, say in a very different habitat, since they are a small population they will evolve very rapidly.[24]

Remember the Founder Effect? It is one of the things that causes speciation.

Let's go back to the Furry Critters. So far, the poor things have endured a change of habitat, an avalanche and an earthquake. A few thousand years after the earthquake we have a large population of silver Furry Critters and a smaller population of gold Furry Critters. This time let's say they had a flood and several individuals from the gold population were washed downstream. They survived (they are good swimmers), but they are now so far from the rest of their group they never make it back. By chance, a couple of the Critters in the new small gold group have a fairly rare gene for long legs. There are so few of them at this point that the gene for long legs can easily spread through the population. This is the Founder Effect again. It's easier for a new variation to spread through a small population than a large one. In a large population, a new mutation will be diluted by all the genes already there. Over time, if the long leg gene has some kind of advantage—the long legs make it easier to run from predators or Lady Furry Critters consider them to be handsome—then the long-leg gene will be able to spread rapidly—rapidly in a geological time scale—through the smaller population.

Eventually most the individuals in the population will have long legs. When they finally leave a fossil or two, they

---

[24] *What Evolution Is*, Ernst Mayr, 2002, p. 175-176

seem to appear suddenly, with no local transitional fossils, completely changed from their ancestors.

Extinction

99.9% (and some say more than that) of all species that have ever lived are now extinct. And it's not even the fault of human beings!

There were five major extinction events in the history of the earth and it is possible that another is going on right now. This time it may be our fault.[25]

No one knows what caused most of the extinction events. It is thought that an asteroid hit the earth at the end of the Cretaceous period about 65 million years ago. The evidence suggests that the resulting global winter killed about 50% of everything alive at that time. It may have caused the death of the dinosaurs (except for a few that were the ancestors of birds) and many large vertebrates.

The largest extinction event was the Triassic-Jurassic Extinction, which killed off about 90% of marine animals and about 70% of land all animals. This extinction happened about 250 million years ago and cleared the way for the age of the dinosaurs. The die-off happened over about a million years, which seems like a long time but is, as you will recall, sudden in geological terms. Many hypotheses have been advanced as to why it happened—gamma rays from a supernova, volcanoes, an asteroid impact—but there's not much evidence to support any of them.

Pseudoextinction (or phyletic extinction) is an interesting idea, actually hopeful in a way. It means that a species can go extinct, but because the population, or part of the

---

[25]Mayr, ibid., 199

Sinosauropteryx prima, thought to be an ancestor of modern birds*

population, evolved into new a species, their genes have not been entirely wiped out and they are not utterly extinct.[26]

In a way, birds represent pseudoextinction of the dinosaurs. The dinosaurs are all gone, but strong evidence suggests that birds evolved from them, so a small remaining trace of once mighty dinosaurs still lives on to this day.

For the most part, evolution proceeds by variation and natural selection. However, if the environment changes too fast or if too few Furry Critters have enough variation in their genome to cope with a rapidly changing environment, extinction is the result.

---

[26]"A Theory of Evolution above the Species Level," Steven M. Stanley, PNAS, February 1, 1975, vol. 72, no. 2 pp. 646-650
*By Sam / Olai Ose / Skjaervoy from Zhangjiagang, China - Dinosaurs!, CC BY-SA 2.0, https://commons.wikimedia.org/w/index.php?curid=4209411

7. Common Descent

Darwin's most important idea was that all living organisms on this planet descended from one or a few common ancestors. That idea is one of the most profound implications of evolution, and is very well supported by the evidence. So what is the evidence? How do we know that is true?

We know today that we share morphology (physical structures) with most vertebrates and genetic similarity with all other animals and even all plants. [27]

Another powerful line of evidence of evolution is revealed by taxonomy.

Taxonomy

Taxonomy is a system of naming all living things. Carolus Linnaeus (1707-78)[28] died 31 years before Darwin was born. He was the first to attempt to classify plants and animals and place them into related groups called taxons. His system is still used today with only minor modifications. Without DNA evidence, using morphology alone, he was able to classify all the living things he knew about. Early on, Linnaeus believed that every single animal and plant was created individually by

---

[27] http://www.scientificamerican.com/article/universal-common-ancestor/ (accessed September 14, 2016)

[28] https://en.wikipedia.org/wiki/Carl_Linnaeus (accessed September 14, 2016)

God. Later he concluded that every genus was specially created, which meant, of course, that species had to evolve.

Linnaeus's method for classifying plants and animals revealed a nested hierarchy. A nested hierarchy is a pattern that looks like the way a tree grows, in fact it is sometimes referred to as the "tree of life."

The major taxons are Domain, Kingdom, Phylum, Class, Order, Family, Genus, and Species

There are several Domain classifications. In fact there are several ways to group the broadest categories of life with only some general agreement. Carl Woese[29] introduced the one I will talk about here. He divided life into three domains: Bacteria, Archaea, and Eukarya. Bacteria are cells that don't have a nucleus. Archaea are single-celled like bacteria and also don't have nuclei but otherwise they are, well, they are way different and are usually found in extreme environments. Eukaryotes are organisms that do have nuclei. Those cells with nuclei tend to "clump" together to form trees, grass, crocodiles and people. Human beings belong to the domain Eukarya.

Originally Linnaeus only recognized two kingdoms, plants and animals, but he didn't have modern scientific equipment or information. Since his time we have added the kingdoms fungi, monera and protista. Since this is the *Pocket Darwin* and not the *Carry-it-Around-in-a-Truck Darwin* I will focus on Linnaeus's original two.

Think of *domains* as base of the trunk of the tree of life. The *kingdoms* are where the trunk of the tree separates into two huge branches: plants and animals.

If you look at a tree you'll notice that there will be one or two trunks that split off and then split off again and again into thousands and thousands of branches. The same is true

---

[29] http://www.igb.illinois.edu/about/archaea (accessed September 14, 2016)

of plant and animal taxonomy. To make things easier I'm going to follow two lines—one for humans and one for roses.

Humans are not plants! So we are in the kingdom animalia with all the other animals. Phyla are the next cluster of branches. Scientists debate the exact number, but there are about 33 animal phyla. At this level plants are usually grouped into *divisions* rather than phyla. Humans are in the phylum *chordata*. That means before you were born you had a cartilage-like rod called a *notochord*. That notochord grew into your spine, which puts you in the subphylum *vertebrata*. It looks like there are a lot of vertebrates around from sharks to naked mole rats, but about 90% of all animals in the world don't have spines and most of those are arthropods like shrimp and spiders.

> **How do we know that?** From Wikipedia "The Angiosperm Phylogeny Group is an international group of systematic botanists who have come together to try to establish a consensus view of the taxonomy of flowering plants in the light of the rapid rise of molecular systematics. The flowering plants (also known as angiosperms, Angiospermae, Anthophyta, Magnoliophyta), are one of the groups of organisms whose classification has been affected most radically as molecular data became available.

Roses are not animals! So they are in the kingdom *planta* and the division *magnoliophyta*. Plants in this division all have flowers. That makes this division a no-brainer for roses.

The next branch up the taxonomic tree is *class*. Humans are in the class *mammalia*. We have hair and sweat glands and drank milk as babies just like lions and tigers and bears (oh, my).

The next branch up for our roses is the class *magnoliopsida*. Plants in this class are also called *dicotyledons*, which means that young seedlings get two little leaves as they grow.

If you take an ordinary pinto bean, stick it into a paper cup of dirt and add a little water, in a couple of days your sprout will have two seed leaves. Kindergarten teachers all over the world have their students do this as a class project. It's fun to watch the shoot come out and open up its two little leaves. Pinto beans are dicotyledons just like oak trees, your grandmother's geraniums—and roses.

As we continue to climb up the taxonomic tree we come next to *order*. Humans are in the order *primata*. That should sound a little familiar. We are primates like all apes and monkeys. With monkeys you are beginning to see some family resemblance. Monkeys, like us, have five fingers witch have flat fingernails. One of the fingers is a very useful thumb. They have eyes in the front of their heads, and a collar bone.

Roses are in the order *rosales* along with stinging nettles and elm trees. Members of this order produce flowers and usually fruits.

We are climbing out into the small, high branches of the tree of life now. The next branch level is *family*. The family that humans fall into is *hominidae*. The great apes fall into this category with us—chimpanzees, bonobos, orangutans and all the various kinds of gorillas. The roses fall into the family *rosaceae*. They produce fruit under their flowers. At this stage you can see that wild roses, which have only five petals, resemble their edible cousins in the family rosaceae, such as apples, strawberries and peaches.

The next branch out is *genus* and here things really get personal. The human genus is *homo*, which means "man." Humans are the only species—species is the next level up—

left in this genus. Once there were a lot of species in the genus homo—*Homo neanderthalus, Homo habilus, Homo erectus*, and so on. I'll go into more detail later when I discuss human ancestry.

The genus for roses is *rosa*. It's difficult to tell how many species of rose there are. Some say as few as 100 or as many as 150. Sometimes it's hard to tell if a rose is its own species or if it's a subspecies, a variety or a subvariety. Roses evolve very freely. Some species of roses are present in the fossil record and are millions of years old, but new roses turn up fairly often. Rose taxonomy at this level is messy but a lot of fun!

Species isn't exactly the last level of taxonomy, but it's as far as I'll go for our purposes. Darwin called his book *Origin of Species* for a reason. A lot of people read the title as if he meant "the origin of all species" as in "the origin of life." Darwin didn't have any way to investigate the origin of life and I doubt he was much interested in it. It's a book about how species evolve to produce new species. Species level is where evolution takes place. It's the only place anything has ever evolved. One phylum can't evolve into another. Only species can do that.

DNA

As you will see from the discussion of the mouse genome, DNA evidence of common ancestry is especially compelling. Everything that is alive shares some DNA similarity. The DNA evidence tracks very closely with the morphological evidence Linnaeus used to make his taxonomy. Not only do we share some of our genes with everything that is alive, but even so-called "junk" DNA—DNA that is not currently in use by our bodies—have similarities and differences.

For example, we humans can't manufacture vitamin C in our bodies and neither can many other primates. When the DNA of vertebrates is compared, we find that both humans and primates have genes for making vitamin C, but the genes for it are non-functional. A mutation destroyed that particular gene some time in our history. This mutation, which blocks the production of vitamin C in our bodies, is identical to the mutation in primates. It's very unlikely that the same mutation occurred independently in both apes and humans. Guinea pigs also can't make vitamin C but the mutation causing it is different from ours. It is clear that the mutation in apes and humans occurred at some time in our past in an ancestor common to us both. It's a family resemblance.[30]

Most people have heard of the Human Genome Project. Scientists made a map of all the human genes (most of them, anyway), which will allow them to eventually figure out what gene does what in the human body. After the human genome was finished, they went to work on several important plants and animals.

A lot of medical research is conducted on mice. All of that research would be useless if mice and men did not have a very great similarity. The publication of the mouse genome[31] revealed that the base sequences of the human and mouse genomes are approximately 80% similar. Our common ancestor was not recent—75 million years ago—and today the differences between humans and mice are obviously very great. But just as obviously, the strong genetic similarities clearly show the existence of our common ancestor.

---

[30]Prediction 2.3: Molecular vestigial characters http://www.talkorigins.org/faqs/comdesc/section2.html#molecular_vestiges (viewed September, 2016)
[31]Waterston et al. in *Nature* 420 (6915): 520-562 Dec 5 2002

Arabidopsis (rockcress) is a genus in the family Brassicaceae

We have between 94-96% of our genome in common with chimpanzees. They are our closest relatives and we are their closest relatives. They are closer to us than to gorillas. About 60% of our genome is shared with fruit flies.[32] We share about 40% of our genes in common with the roundworm Caenorhabditis elegans.[33] About 20% of human disease genes have counterparts in yeast.[34] We share about 33% of our genetic material with Arabidopsis[35] a tiny mustard plant, which is more closely related to cabbage and kale than to us, but nevertheless bears a family resemblance in its genome.

---

[32]NHGRI, Comparative Genomics, https://www.genome.gov/10005835/background-on-comparative-genomic-analysis/ (viewed September, 2016)
[33]http://web.ornl.gov/sci/techresources/Human_Genome/publicat/hgn/v10n1/hgn101_2.pdf, p9 (viewed September 2016)
[34]http://www.yeastgenome.org/yeast-are-people-too (viewed September 2016)
[35]http://www.ncbi.nlm.nih.gov/pmc/articles/PMC1283769/ (viewed September 2016)

8. Human Evolution

The Missing Link: Human or Ape?

"The Missing Link" is not a phrase that scientists use. It has been a great favorite with newspaper reporters and the media for more than a century, but there is no one link between humans and apes that is or ever was missing.

The first Neanderthal skull was found in 1829 and another was found in 1848 but they didn't get any attention. In 1859 quarry workers found several primitive fossil bones in the Neander valley in Germany. They gave the bones to a naturalist named Johann Karl Fuhlrott, who, with anatomist Hermann Schaafhausen, announced the find a year later. In 1864 anatomist William King named this find *Homo neanderthalis*. These were only a skull cap and some leg and arm bones. In 1887 two complete skeletons were found in Belgium.

At first it was thought that Neanderthal was simply an elderly, deformed human with some kind of crippling disease. As more fossils were found it became obvious that couldn't be true. Now scientists have bones from about 400 individuals and it's clear they are a species in their own right. They have several features distinct from humans especially in their face and skull. Early on, the press tagged them with the label the "missing link." They aren't. DNA testing indicates we share a common ancestor with them, probably *Homo heidelbergensis* who lived about 120,000 years ago.[36]

---

[36] http://anthro.palomar.edu/homo2/mod_homo_2.htm (viewed September 2016)

The earliest ape-like fossils date back to more than 15 million years ago. They were clearly not yet even slightly human.

Paleoanthropologists, the people who study human evolution, refer to humans, chimpanzees and their direct fossil ancestors as "hominids" or "hominini."

The earliest hominid may have been Sahelanthropus tchadensis who lived around seven million years ago. This creature probably did not walk upright and had a small, chimpanzee-like brain. However it did have a few features that are only found in human ancestors.[37]

Australopithecus: The Taung Child and Lucy

In 1925 Raymond Dart found the skull of a hominid species he called *Australopithecus africanus*, which lived about 2.5 million years ago. It was about three or four years old at the time of death. Dart nicknamed it the Taung Child after the place where it was found in South Africa. It had a very small brain, even for a child. Dart's discovery was not well-received at first. Not only did it contradict the "big-brain-first" hypothesis, but there was still some resistance to the idea that humans evolved in Africa. Yet that little skull was clearly a relative of some kind. How could they tell?

If you are reading this you are almost certainly in the genus "homo" and the species "sapiens." All creatures with a skull have a foramen magnum. That's a hole in the skull that allows the vertebrae to attach to your head. Yours is at the base of your skull and that is a consequence of the fact that you walk upright. Chimps have theirs higher up on the back of the skull. They can walk upright for a while if they want to, but they are not *obliged* to walk upright like we are. Cats and dogs, almost can't walk upright at all. Their fora-

---

[37] http://humanorigins.si.edu/evidence/human-fossils/species/sahelanthropus-tchadensis (viewed September 2016)

men magnum is in the back of their skull. In order to tell if something walked upright all the time or even some of the time all you need to look at is the skull.

In spite of its very small brain, the position of the Taung Child's foramen magnum tells us that it walked upright. Of all the apes alive today, only humans walk upright and that is why it is considered to be a uniquely human trait. The Taung Child's genus—*Australopithecus*—was probably not directly ancestral to humans, they were a side branch off the family tree.[38]

Donald Johanson discovered another *Australopithecus*, which he gave the species name "afarensis," in eastern Africa in 1974. The fossil was nearly 40% complete and Johanson nicknamed it "Lucy." Lucy lived a little less than 4 million years ago. Her species walked upright but with a slight stoop. Since she was discovered, remains from about 300 individuals of her species have been found. *A. afarensis* were obligatory upright walkers, like we are, but their hands and shoulders still had some adaptive traits for living in trees. They were small—about 3 feet tall—and barrel chested. They had a very small brain. Yet they had many human-like traits. For example, their hips and their teeth were much more like a human's than an ape's. They are nearly a perfect transitional fossil with some ape characteristics and some human characteristics.[39]

Homo Erectus and Java Man

Eugene Dubois, believing that ancient man would be found in Asia rather than Africa, went to Java and in 1891 he found a hominid skullcap, a leg bone and some teeth. The skull indicated a bigger brain than an ape but it was a lot smaller than a modern human or a Neanderthal. Dubois named his find "Java Man."

---

[38] *Wisdom of the Bones* by Alan Walker and Pat Shipman, 1997, p.90
[39] *Lucy, The Beginnings of Humankind* by Donald Johanson, Maitland Edey, 1990

Once again the press called it the missing link. It wasn't. The femur is probably modern human, but the skullcap was *Homo erectus*. From 1931 to 1941 Davidson Black, G. H. R. von Koenigswald and others found several more specimens of *Homo erectus* in China. In 1983 Kamoya Kimeu discovered a nearly complete *Homo erectus* skeleton which was nicknamed "Turkana Boy" after the lake near where it was found. *H. erectus* was tall, almost six feet, and there's strong evidence that it made and used tools and could use fire as far back as 1.5 million years ago. It probably died out about 500,000 years ago. Before disappearing, though, it traveled out of Africa and all over much of the world which is how 19th and early 20th century fossil hunters could find it in China and Indonesia.[40]

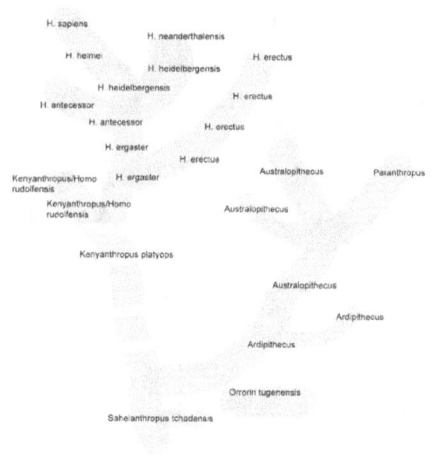

The Human Family Tree

The family tree of *Homo sapiens* is not a finished picture. The "leaves" on this tree may not stay in the same position as they are now. The arrangement keeps getting tweaked as new fossils are discovered. Like the Neanderthal, who turned out to be a cousin rather than an ancestor, *Homo erectus* could possibly lose its ancestral standing. Even so, the general trend from very primitive and apelike to more and more modern-looking is very clear without knowing all the fine details.

---

[40] *Wisdom of the Bones*, ibid, p42-45

The hominid fossil record documents the evolution of humans from an ape-like ancestor through a series of stages involving the acquisition of bipedal walking, smaller teeth, enlarged brain capacity, tool making capabilities and various aspects of culture. The most recent common ancestor of modern apes and human beings lived some time between 5 and 10 million years ago. This information is deduced from molecular data, but the fossils bear out that deduction.

- The earliest fossil hominid or near-hominid, *Sahelanthropus tchadensis*, lived between 6 and 7 million years ago. Its cranial capacity was very small, only about 350 cc. It is not known whether it was bipedal.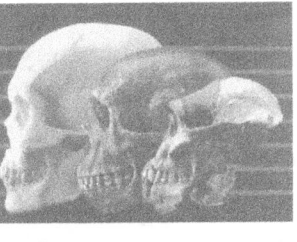
- *Ardipithecus ramidus* is between 5 and 6 million years old.
- *Australopithecus afarensis*. Though there are many examples of this species, the first discovered and most famous is called "Lucy." *A. afarensis* lived between 3 and 4 million years ago, was bipedal, though it walked a little stooped over, and had a cranial capacity from about 375 to 550 cc.
- *Kenyanthropus platyops* is about 3.5 million years old.
- *Australopithecus africanus* is between 2 and 3 million years old. Their teeth are very similar to those of humans but their cranial capacity is between 420 and 500 cc, which is a little larger than chimp brains.
- *Australopithecus aethiopicus* which is not directly ancestral to humans, lived between 2.6 and 2.3 million years ago and had a cranial capacity of about 410 cc.
- *Australopithecus robustus* is also not directly ancestral to humans and lived between 2 and 1.5 million years ago. It

may have used digging tools and had a cranial capacity of about 530 cc.
- *Homo habilis* is so called because of evidence that they made and used stone tools. Their cranial capacity was between 500 and 800 cc and there is some evidence that they were capable of rudimentary speech.
- *Homo ergaster* is an early erectus-like hominid which lived about 1.8 million years ago. Its cranial capacity varied from 600 to 680 cc.
- *Homo erectus* lived 1.8 million to 300,000 years ago. Its cranial capacity was about 750 to 1225 cc. *H. erectus* made sophisticated stone tools and probably used fire.
- *Homo heidelbergensis* first appeared about 500,000 years ago. The brain size is larger than *H. erectus* but smaller than modern humans, averaging about 1200 cc.
- *Homo neanderthalensis* existed between 230,000 and 30,000 years ago. Their brain size was slightly larger than that of modern humans at about 1450 cc. They made sophisticated stone tools, were formidable hunters and are the first people known to have buried their dead.
- *Homo sapiens*, modern humans, appear about 120,000 years ago. Modern humans have an average brain size of about 1350 cc.

9. The Fossil Record

Only a small portion of the world has been geologically explored. Only organic beings of certain classes can be preserved in a fossil condition, at least in any great number. Widely ranging species vary most, and varieties are often at first local, — both causes rendering the discovery of intermediate links less likely. Local varieties will not spread into other and distant regions until they are considerably modified and improved; and when they do spread, if discovered in a geological formation, they will appear as if suddenly created there, and will be simply classed as new species.[41]

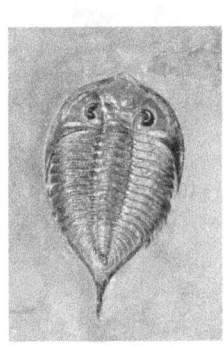

Silurian Trilobite

In the 1700s and early 1800s, geologists and engineers observed that rock and soil are arranged in layers. They assumed, as we do now, that deeper meant older. They also noticed that fossils were distributed in the layers in a way that was not random. Some kinds of fossils were always found above or below other kinds. They eventually saw that this order was consistent everywhere in the world. This order has an undeniable general trend. As the fossils become more recent in time, the more they resemble modern forms.

---

[41] *Origin of Species*, ibid., Ch. 14

Though it is far from complete, the fossil record is one of the most powerful and persuasive ways we can examine the history of life on earth, investigate the details of the "tree of life" and trace the ancestry of currently living organisms, including human beings.

About 600 million years ago, the imprints of soft-bodied animals appear in the fossil record. This is called the Ediacaran or Vendian era of earth's history. Not a lot is known about these early creatures, though quite a few of them have been identified, mostly algae and sponges. Some of them are believed to be ancestral to the animals and plants found in the Cambrian Era.

Archaeopteryx

The evolution of life in the Cambrian Era is sometimes referred to as the "Cambrian Explosion." This "explosion" took place over about 50 million years, making it just about the slowest explosion in the history of the universe. Nevertheless, after more than 3 billion years of blue-green algae, suddenly—at least in geological terms—more complex animals began to show up in the fossil record. Most of the modern phyla appear for the first time in the Cambrian, including our own, Chordata. Chordata was represented by a tiny little creature about an inch long. We chordates have come a long way in 500 million years.

All plants and animals that lived during the Cambrian Era lived in the sea. The earth was barren and rocky. By about 500-450 million years ago fish-like animals dominated the waters. They were the first vertebrates. They didn't look too much like modern fish. They had armor instead of scales and badly-formed fins. During this time, insects and plants began to spread out on to the land. Such a rich food source won't go long before something evolves to take advantage of it. By about 380 million years ago the fish-like creatures also began to move onto land.[42]

Scientists have found numerous fossils which document each of these stages of development. Remember, even 250 years ago geologists figured that deeper meant older. The strange armored fishlike fossils are always found in older geological strata than an animal like Tiktaalic[43], which has some features of fish and some features of a four-legged land animal. More modern looking four-legged animals are always found above transitional fossils like Tiktaalic.

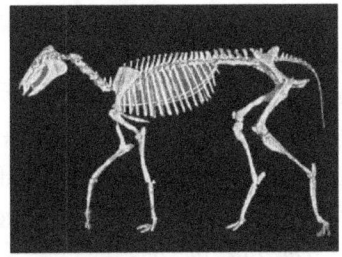

Miohippus an ancestor of modern horses which lived about 32-25 million years ago.

Simpler organisms such as bacteria appear long before protists (more complex single-celled forms), which in turn appear long before multicellular plants and animals. Each group of multicellular organisms also appears in a definite general order. Thus within the vertebrates, fishes appear first, then amphibians, then reptiles, then mammals and then birds. Within the mammals, apes appear before hominids and hominids before modern humans.

Even if almost no fossils had ever been found, the general trend of evolution would have been very clear.

---

[42] *History of Life*, Richard Cowen, 1995, p. 64
[43] http://tiktaalik.uchicago.edu (accessed September 2016)

Transitional Fossils

> "The supposed lack of intermediary forms in the fossil record remains the fundamental canard of current antievolutionists. Such transitional forms are scarce, to be sure, and for two sets of reasons—geological (the gappiness of the fossil record) and biological (the episodic nature of evolutionary change, including patterns of punctuated equilibrium and transition within small populations of limited geological extent). But paleontologists have discovered several superb examples of intermediary forms and sequences, more than enough to convince any fair-minded skeptic about the reality of life's physical genealogy."[44]

Strictly speaking, every fossil is a transitional form, since every fossil organism has an ancestry and every fossil has descendants unless it is the last in an extinct line.

Birds

In Darwin's time the fossil record was very poor. The dinosaur-bird transitional fossil *Archaeopteryx* was not discovered until 1861, two years after the publication of the *Origin of Species*.

When Archaeopteryx was discovered, the paleontologists at the time thought it was just a fossil of a small dinosaur. It was not until 1863 when the fossil was being prepared for display in the museum that the feather impressions were found. "Archy" is a perfect transitional fossil. He has teeth, a bony tail and claws on the ends of his wings. But he also has feathers and a bird-like breast bone that looks like it would have allowed him to fly a little, maybe about as much as a chicken. Archaeopteryx is not ancestral to modern birds—there are fossils of true birds that are older than he is—but he is clearly a cousin off the main line of descent.

---

[44] Stephen Jay Gould, "Hooking Leviathan by Its Past," 1994; in *Dinosaur in a Haystack: Reflections in Natural History*.

Even as late as the 1980s scientists were in doubt about whether birds evolved from actual dinosaurs or if they evolved from some other small reptile. The discovery, in the 1990s, of fossil dinosaurs much older than Archaeopteryx with clear feather impressions clinched the deal. Researchers at the University of Washington added more data to the pile when they demonstrated that birds share a unique bone structure with a group of small dinosaurs called *coelurosaurs*.[45]

There are still a few holdouts who think birds evolved from some other group of Jurassic animals even though the evidence against them is overwhelming. But nothing in science is ever finished or completely settled. They could still be right and they could prove it with the right fossils and more data.

Horses

Horse evolution has been well documented by many hundreds of transitional fossils for more than a hundred years. In the 19th century, O. C. Marsh discovered a large number of horse fossils in various parts of North America. Marsh was a famous fossil hunter who discovered hundreds of dinosaurs, flying reptiles and even *Ichthyornis* a bird from the Cretaceous period that looked like a seagull except it had teeth.

Horses originally had fairly flexible legs and many toes. They were also very small, no larger than a medium-sized dog. As their habitat became drier and more grassy they were forced to evolve to meet the new environmental challenges. Their teeth changed from shapes suitable to eating leaves and fruits to something more suitable to grinding grass and

---

[45] http://www.washington.edu/news/2000/08/09/microscopic-bone-evidence-supports-dinosaur-bird-evolution-link (accessed September 2016)

twigs. Out on the grassy plains where they were easy pickings for predators, the faster horses were the best survivors. Fairly rapidly—over only 7 or 8 million years, horse's legs became longer and stronger and horse's toes evolved into a hoof, all of which made them much faster runners. There are transitional fossils documenting hoof evolution almost every step of the way. [46]

Whales

For a long time there wasn't much in the fossil record about whales. Whales are mammals just like people, dogs, and lions. Their babies drink milk, they have hair and as embryos they develop legs, which are usually reabsorbed before they are born. In fact whales are still sometimes born with small vestigial legs. Therefore scientists have always believed that whales descended from some kind of land-dwelling mammal. Modern DNA evidence has revealed that they are correct and whales are related to hippos. Both the DNA evidence and the fossil evidence support the idea that whales evolved from an animal that lived on land to one that lives entirely in the water. Until the 1990s there were only a few scraps of fossil bone to document whale evolution but now there are many transitional fossils showing the evolutionary journey from land to sea.

Modern whales have very unique ear structures and that has been very helpful in tracing their ancestry. *Pakicetids* have just such an ear structure and they were about the size and shape of dogs and lived on the seashore. *Pakicetus* also had teeth like later fossil whales but not like modern whales.

Amblocetus was sort of like a sea lion. It had legs but it lived at least part of the time in the water near the shore and in swamps. It had that same unique ear structure as modern

---

[46] http://www.talkorigins.org/faqs/horses/horse_evol.html (viewed September 2016)

whales and also had similar teeth and some parts of its skull are like those of modern whales.

Next in line in whale evolution are *Remingtonocetus, Protocetus, Dorudon,* and *Basilosaurus* I will skip over all but *Basilosaurus. Dorudon* and *Basilosaurus* were fully aquatic. They still had tiny, well-formed legs but they spent all their time in the water and looked a lot more like modern whales with many modern whale features. A famous *Basilousaurus* fossil was found in the Egyptian desert but they have also been found in Louisiana, Mississippi and Alabama. [47]

The Bushy Tree

In my descriptions of the evolution of the bird, horse and whale, I have made them seem very straightforward. We traced a straight line from (more or less) the beginning to the final product we see today. Evolution never works that simply. Remember the "Tree of Life"? The branches fork over and over and over creating a tangled maze that sometimes isn't easy to untangle. Paleontologists, anatomists and zoologists are always working to find the correct layout of the tree. The lineages I have presented here are generally correct, but the details aren't finished. There are always new fossils to find and always more to know!

Living Fossils

So-called "living fossils" are plants or animals that have no close living relatives except in the fossil record. The gingko tree has only one member in its class, its order, its family, its genus and there is only one species. Fossils of the gingko tree go back 170 million years. It has evolved a little, but not much in all that time.

---

[47] http://evolution.berkeley.edu/evolibrary/article/evograms_03 (viewed September 2016)

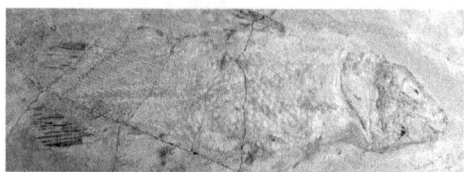

Coelacanth in the fossil record

A modern coelacanth

The coelacanth is a fish related to the first land vertebrates. They were thought to have gone extinct 80 million years ago. In 1938 a living coelacanth was caught off the coast of South Africa, causing enormous excitement among scientists. Modern coelacanths are not identical to their ancient relatives, like the gingko they have evolved to some extent, but they are clearly descended from a line that we had previously know only from fossils.

Ginko then ...   ... Ginko now

10. The Age of the Earth

The universe is about 12-15 billion years old. The earth is about 4.5 billion years old. There are now many methods for establishing those ages, but that was not always the case.

> ". . . we find oysters together in very large families, among which some may be seen with their shells still joined together, indicating that they were left there by the sea and that they were still living when the strait of Gibraltar was cut through. In the mountains of Parma and Piacenza multitudes of shells and corals with holes may be seen still sticking to the rocks . . ."[48]

Long before Darwin, geologists and perceptive observers knew that the earth was ancient. How ancient was very much in doubt. In the 1700s, early geologists were estimating the age of the earth from 75,000 years to several billion years, and a century later estimates were not much more accurate.

In 1862 William Thompson, who became Lord Kelvin because of his brilliant work in physics, estimated the age of the earth at 98 million years, based on the temperature of the earth. In 1897 he revised his estimate downwards to 20-40 million years. Lord Kelvin didn't know about radioactivity, which keeps the earth warm. This was very disappointing to people researching the new science of evolution. Forty million years—even 98 million—was too little time for life

---

[48] Leonardo Da Vinci, *Selections from the Notebooks of Leonardo Da Vinci*, 1510.

on earth to have evolved as Darwin proposed. If Kelvin's estimates had been true, Darwinian evolution would have been disproved.

Radioactivity was discovered around the turn of the 20th century. In 1905, Bertram Boltwood discovered that uranium disintegrates into lead, which means the older a mineral with uranium in it, the more lead it should have. Studying samples whose relative geologic ages were known, Boltwood found that the ratio of lead to uranium did indeed increase with age. After estimating the rate of radioactive change in lead he calculated that the absolute ages of his samples to be from 410,000,000 to 2,200,000,000 years old. Later study would show that his figures were too high by about 20%, but it was enough to prove Lord Kelvin was wrong.[49]

How Do We Know the Earth is Old?

Radioactivity was discovered in 1896 and in the 1950s geologists began to use radiometric dating to find out the age of rocks. Radiometric dating measures the decay of radioactive elements. There are about 40 different radiometric dating methods, but I'll just explain one.

First of all you have to find rocks that were formed when lava cooled. Those are called igneous rocks. Potassium is very abundant on the earth and those rocks will have potassium in them. Potassium-40 is radioactive and decays into calcium-40 and argon-40. Those are called daughter elements.

Argon is a gas. When the rock is heated to molten lava, almost all the argon escapes. When the rock cools and hardens the potassium-40 is still radioactively decaying but now the argon-40 can't escape. Then all you have to do is measure

---

[49] Much of this is from "Changing Views of the History of the Earth" by Richard Harter, used with permission http://www.talkorigins.org/faqs/geohist.html (viewed September 2016)

the amount of parent and daughter elements and calculate a ratio that will tell you how long the potassium-40 has been decaying since the last time it was cooked in a volcano or in the earth's magma.

Note above I said *almost* all the argon escapes from the molten lava. Scientists assume some argon was in the rock when it cooled. So dates are never expressed exactly. They are always stated as a range like "3.62±0.13" which means "3.62 billion plus or minus .13 billion years."

Radiocarbon Dating

Radiometric dating is sometimes confused with *radiocarbon* dating, which is really different. Radiocarbon dating measures the amount of carbon-14 in something that was once alive. This kind of dating is used a lot in archaeology. It can be used to date wood, bone, paper, fabric—anything that was alive sometime in the past.

While you live you are ingesting a little bit of new carbon-14 all the time. You breathe it from the air and you eat it in food. When you die you don't take any more in and whatever carbon-14 was present begins to decay away. While that's going on, your bones (and your clothes) can be dated using this method. But after about 45,000 years or so all the carbon-14 is gone and there's no way to tell how long it has been gone. You can't date fossils with it because, though they were once alive, they are either vastly too old to have any carbon-14 left or they have become mineralized and have turned into stone.

11. The Beginning of Life

Abiogenesis, the study of how life began, is not actually a part of the Theory of Evolution. Though it is clear that life must have begun some way, exactly how life began is still unknown. Evolution did not begin until life existed and could replicate itself.

When the earth first formed more than 4.5 billion years ago, the crust was hot and geologically very active. No life could form until it cooled off and settled down. That took about a billion years.

The oldest fossils ever discovered are bacteria-like organisms called cyanobacteria or "blue-green" algae. Those tiny fossils date back to 3.5 billion years ago. They formed within a few million years after the crust of the earth cooled enough to allow life to form. It is astonishing how quickly life took hold. It's almost like the moment it was possible, life sprang into being.

What is life?

According the Harvard biology professor Andrew Knoll, one definition of life is "something that can make more of itself." In order for evolution to begin, something had to be able to make more of itself and do it *imperfectly*. That imperfection is very important because that becomes the variation that natural selections needs in order to "select."

If you follow Knoll's definition, viruses are alive. Other scientists don't think they are. Yes, they make more of themselves but they need a living cell to do it. They do evolve, though. The HIV virus evolves so quickly almost anyone who is infected has a virtually unique version of the virus. Even though viruses are in the gray area between life and non-life, some researchers think that early life may have been much like a virus.

But cyanobacteria and even viruses are pretty far down the road headed toward life. True first life would have been just some self-replicating molecules. How did something like that get started? That is the big question. Several answers have been proposed.

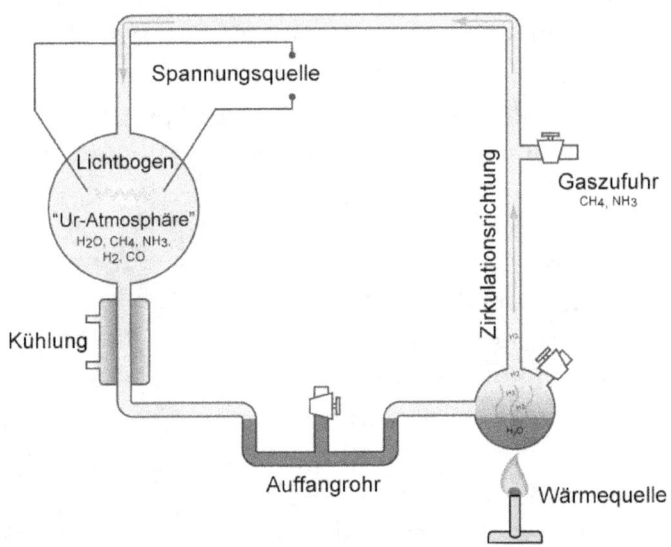

The Miller-Urey Experiment

In the 1920s Aleksandr Oparin and J. B. S. Haldane hypothesized that simple organic molecules like sugars and amino acids could form spontaneously under certain circumstances. In the 1950s Stanley Miller and Harold Urey tested that hypothesis.[50] They designed an apparatus that simulated what they thought might be the atmospheric conditions on early earth. They shot electricity through the apparatus to simulate lightning. A week later they found that amino acids and other organic molecules had formed. Since that time, many similar experiments using different combinations of gases have produced a wide variety of organic molecules, including the nucleotide bases of RNA and DNA.[51]

The Primordial Organic Soup

Oparin, in the 1930s, envisioned that the organic molecules would, over vast spans of time, accumulate in the shallow seas to form a "sea of organic soup." Under such conditions he thought that smaller organic molecules (monomers) would combine to form larger ones (polymers). Based on evidence gathered since Oparin's time, a lot of scientists think it might be more likely that organic polymers formed and accumulated on rock or clay surfaces rather than in the primordial seas.

Hydrothermal Vents

Another hypothesis is that the origin of life occurred around hydrothermal vents. The action of plate tectonics causes cracks in the ocean floor. Water pours down into those cracks, gets superheated and then spews back out again. When the hot water shoots back up to the ocean it collects minerals such as sulfur, iron and hydrogen. Very prim-

Günter Wächtershäuser proposed that life might have originated in hydrothermal vents.

---

[50] https://en.wikipedia.org/wiki/Miller–Urey_experiment (viewed September 2016)
[51] http://www.livescience.com/1804-greatest-mysteries-life-arise-earth.html (viewed September, 2016)

itive single-celled organisms swim in that hot water and feed on those chemicals. Some researchers believe it's possible first life could have safely formed in the iron sulfide that gets deposited around hydrothermal vents.[52]

After the first organic molecules form, how could they have assembled spontaneously into more complex structures? That is still unknown, but scientists have synthesized several different molecules, precursors to life, which are called "protobionts." They have been able to make protobionts that resemble living cells in several ways, helping us to figure out how aggregations of complex nonliving molecules became living cells.[53]

---

[52] https://en.wikipedia.org/wiki/Iron%E2%80%93sulfur_world_hypothesis (viewed September 2016)
[53] http://www.biocab.org/protobiont.html (viewed September 2016)

12. God and Darwin

It is often said that you can't believe in evolution and believe in God at the same time, or that you can't be a Christian and believe in evolution. Nothing could be further from the truth. You can't be a Biblical literalist and believe in evolution, but only a minority of Christians believe that the Bible is literal history. All you have to believe is that God is in charge of evolution, that God made the universe and everything in it and evolution is how he created life on earth.

The essential message in Genesis is "I made you." Shepherds of 3000 years ago did not need to know the details of biochemistry or fossilization. What would they have done with such knowledge?

An omnipotent God could have created the raw material of the Big Bang and touched it off. An eternal God would not have considered 14 billion years to be a long time. Such a time-span would be a mere eye-blink. An omniscient God would have known in advance that a creature—perhaps many creatures on many planets—would eventually evolve that could detect his presence and form relationships with him.

Such a God would have built randomness into the system to ensure that all created beings would have true free will and could make true moral choices.

Evolution does not speak to any of these issues. It does not contradict the existence of a God or preclude a belief in creation. For Christians and other religious people who believe the evidence for Darwin's theory is compelling, the study of evolution and the natural sciences is a study of God's handiwork.

13. BONUS ESSAYS

Two Opposable Thumbs Up for Evolution!

Okay, show of hands. Does everyone reading this have a spine? Yes? Pretty much everyone? Great! Hold that thought.

I get asked a lot "what is the evidence for evolution?" People ask me at parties, casual get-togethers, sometimes even in parking lots when someone has spotted my bumper stickers (Two opposable thumbs up for evolution!). It's a darn big subject to condense down into a paragraph. I'm still working on it. But what I generally say is "you are a mammal." So far only one person has said "I am not!"

Humans are animals. There's just no getting around it. We aren't plants. We aren't fungi. There are some other kingdoms, but nobody reading this is single-celled. In any case almost all of us belong in the kingdom animalia along with all the other animals.

We are in the phylum chordata. So when you were a fetus you had a notochord, a rod made out of something similar to cartilage. As you developed, that notochord grew into a spinal cord and vertebrae. If you answered "yes" to my question about your spine, those vertebrae put you in the subphylum vertebrata, along with fish, amphibians, snakes and birds.

Feel your hand. Hopefully it's warm, not cold like a snake. Do you have feathers? No? Do or did you have teeth? Yes? Good! You aren't a bird! Most people mostly have hair. I'm guessing you drank milk as a baby. You are in the class mammalia, along with dogs, dolphins, mice and elephants.

Okay, you're a mammal. That's still a long way from being human. As far as we know you're some blind mole that lives out in the desert. Do you lay eggs? Do you nurture your young in a pouch? If you answered both those questions "no," then you are in the subclass eutheria.

Now for the easy stuff—or the difficult stuff, depending on your point of view. You should have five fingers and five toes on each hand and foot. One of those fingers should be a thumb. You have flat fingernails on those fingers and toes and in addition to that you have a collarbone. All of that puts you in the order primata (where the monkeys are).

Okay, we're almost there. Tail? Nope. Well, not any more. You did have one as an embryo. Do you walk on two legs? Yup. Your spinal cord and vertebrae are "S" shaped. That puts you in the family hominidae.

You have a big brain, small teeth and a flat face. That puts you in the genus homo. And finally, you are a member of the species sapiens. There used to be many species in the genus homo but we are all that remain.

And so I tell people when they ask me, the evidence for evolution is in front of you every time you look in the mirror.

Obviously there's lots of other evidence: DNA, geographical species radiation, ring species, tons of fossil evidence and that little detail that speciation has been observed both in the wild and in the lab.

Why on earth would someone reject evolution? Nobody rejects scientific theories of reproduction in favor of the stork theory, what's the big deal about being a mammal?

When Darwin was born in 1809, naturalists already had noticed that species change through time. Geologists had already decided the earth had to be very ancient. They'd seen the fossils in the various geological strata. In fact they had already noticed that some fossils were found only in certain geological layers. They assumed deeper layers meant older—and as it turns out they were right—and they were already using those index fossils to tell which geological layer they were dealing with. They knew the fossils they saw in deep geological layers were of various organisms unlike anything alive today.

And those geologists and naturalists were all creationists. Most of them believed in a literal reading of Genesis. How could they harmonize what they saw with what they believed? Generally they believed in successive creations. They decided that there had been many creation events through time and the one recounted in the Bible was only the most recent.

When Darwin set sail on the Beagle, he had a fresh ministerial degree and he had every intention of returning in three years (instead of the five the voyage actually took), and settling down to be a country pastor who collected beetles in his spare time. It didn't work out that way. Sometime in 1837 or '38 he wrote in his diary "In July opened first notebook on transmutation of species. Had been greatly struck from about the previous March on character of South American fossils, and species on Galapagos Archipelago. These facts (especially latter), origin of all my views."

Intelligent Design is a product of evolution. In 1987 the Supreme Court ruled in *Edwards v. Aguillard*, that creation science was a religious apologetic and not science and couldn't be taught in science classes.

In 2005 during the trial *Tammy Kitzmiller, et al. v. Dover Area School District, et al.* Barbara Forrest pointed out that the book *Of Pandas and People* was being revised in 1987. *Pandas* was a creationist textbook intended for science classes. I'm given to understand it is now in wide use with homeschoolers and in private Christian schools. Forrest pointed out in the early drafts of the book there were no references to Intelligent Design or the Designer. It was strictly focused on creation science. But in 1987 *Pandas* evolved. All references to "creation," "creator" and "God" were removed and replaced with "design," "designer" and "intelligent designer."

What selection pressure caused that evolution? *Edwards v. Aguillard*, of course. The court wanted to make it clear that scientific arguments were always welcome in public schools. They wrote:

> "We do not imply that a legislature could never require that scientific critiques of prevailing scientific theories be taught. . . . [T]eaching a variety of scientific theories about the origins of humankind to schoolchildren might be validly done with the clear secular intent of enhancing the effectiveness of science instruction."

So creationism had to evolve or die. Intelligent Design, its proponents tell us, unlike creation science isn't about creationism. It's science. There's scientific research to back it up. Lots and lots of scientific evidence to support it. There's irreducibly complex bacterial flagella and blood clotting cascades. There are explanatory filters and specified complexity. Gone were the quaint, old fashioned creationist arguments about moon dust proving the earth is young, gone were the arguments about hydrological sorting of fossils during the Great Flood.

Also gone was the argument about Bomby the Bombardier beetle. This was one of my favorites. The Bombardier beetle, so the argument goes, defends itself by shooting out a boiling hot toxic spray from its posterior. Duane Gish's version claims that hydrogen peroxide and hydroquinones would explode spontaneously if mixed without a chemical inhibitor, and that the beetle starts with a mix of all three and adds an anti-inhibitor when he wants the explosion. The creationist reasoning goes that since none of the process would work without all the parts being present at the same time, it all had to be created together at once. It couldn't have evolved. That's not very accurate. Hydrogen peroxide and hydroquinones will not explode if mixed together. After more than 40 years of being corrected, Duane Gish died in 2013 and never changed his story.

The evolution of the bombardier beetle is very well understood. There are lots of kinds of beetles who do their chemical thing a little differently, many with fewer components. You don't need all the parts all at once to make it work. The reason I bring it up now is because recently, a clueless creationist, not up to speed with the latest round of arguments, mentioned it on a debate e-list. It suddenly dawned on me that this was an early primitive fossil of the argument about irreducible complexity. Michael Behe's argument about the bacterial flagella descended from Gish with modification.

Intelligent Design began its life as an argument for the existence of God.

See if this sounds familiar:

> "I understand [Intelligent Design] to be a reformulation of an old theological argument for the existence of God, an argument

that unfolds in the form of a syllogism, the major premise of which is wherever there is complex design, there has to be some intelligent designer. The minor premise is that nature exhibits complex design. The conclusion, therefore, is nature must have an intelligent designer."

That is a quote from John F. Haught, professor of theology at Georgetown University and author of *God After Darwin*, and *Deeper Than Darwin*. It was part of his testimony at the Dover trial.

When asked if he could trace the antecedents for this argument, he went on to say:

"Well, two landmarks are Thomas Aquinas and William Paley. Thomas Aquinas . . . formulated what are called the five ways to prove the existence of God, one of which was to argue from the design and complexity and order and pattern in the universe to the existence of an ultimate intelligent designer. . . . Thomas Aquinas ended every one of his five arguments by saying that this being, this ultimate, everyone understands to be God.

And William Paley, in the late 18th and early 19th Century, is famous for formulating the watchmaker argument, according to which, just as you open up a watch and find intricate design [which] lead[s] you to postulate the existence of a watchmaker, so also the intricate design and pattern in nature should lead one to posit the existence of an intelligent being . . . .

And like Aquinas, William Paley also said to the effect that everyone understands this to be . . . the creator God of biblical religion."

"Okay," Intelligent Design apologists tell me. "So what? Just because a scientific idea has religious implications, doesn't mean it's not science."

And they are absolutely correct.

The Big Bang hypothesis met with a fair amount of resistance in the beginning, partly because it sounded religious. One of its early proponents was a Jesuit priest named Georges Lemaître. He was a scientist and an astrophysicist but, still, you knew he liked it because it looks like the creation event.

For decades, creationist Hugh Ross has been infuriating young earth creationists with his reverence for the Big Bang. In one of his essays, he writes:

> [T]he Bible indirectly argues for a big bang universe by stating that the laws of thermodynamics, gravity, and electromagnetism have universally operated throughout the universe since the cosmic creation event itself. In Romans 8 we are told that the entire creation has been subjected to the law of decay (the second law of thermodynamics). This law in the context of an expanding universe establishes that the cosmos was much hotter in the past. In Genesis 1 and in many places throughout Job, Psalms, and Proverbs we are informed that stars have existed since the early times of creation.

And you know what? It's legal to teach the Big Bang in science classes. Its religious implications are not a hindrance to it because of all the scientific evidence supporting it. It can't be barred from public schools even though millions of people believe the Big Bang is the original creation event and therefore an argument for the existence of God.

What's up with that? How did it get accepted by scientists and into public schools? Early on, scientists resisted it; the steady state universe was more appealing. But, darn it, galaxies are red-shifted, indicating that the universe is expanding. There are detailed measurements of the cosmic microwave background, primordial elements are abundant. All that stuff supports the Big Bang and these days most cosmologists are convinced that it happened. The wall of resistance was beaten down with evidence. Before it was taught in high school science classes, decades of research was carried out. Thousands of experiments were conducted. Hundreds of papers were written. Conferences were held. Arguments were batted back and forth. Over time the evidence piled up and few scientists in the field could resist it even though it had a clear religious implication.

Compare all that to intelligent design. I'm sure some of you have read the Wedge Document. If you haven't read it, it's in thousands of places all over the web. You can find it here. (pdf) https://ncse.com/files/pub/creationism/The_Wedge_Strategy.pdf At first the Discovery Institute denied they wrote it but eventually came clean about it and put it on their website. They still try to deny it's an actual game plan, but it's pretty clear that the plan set out in that document is pretty much how Intelligent Design advocates have proceeded.

The Wedge document says in part:

> "Design theory promises to reverse the stifling dominance of the materialist worldview, and to replace it with a science consonant with Christian and theistic convictions."

One of the first goals it sets out is writing and research. Admirable goals. They mention that "Without solid scholarship, research and argument, the project would be just another attempt to indoctrinate instead of persuade." I agree.

The Wedge strategy is in three phases. Phase I is about research and writing papers. Phase II prepares the public for the good news about ID. Phase III is about getting ID into colleges and scientific settings. Phases I and III languish. Their goals have largely not been met. Yes, they have produced some popular writings on Intelligent Design, they have gotten a couple of papers published. I think they've even presented at a conference or two. But to this day other than Dembski's mathematical odds calculations and Behe's flagellum, no scientific research has been conducted. Most of the work with a somewhat scientific flavor has been criticisms of evolution, not the gathering of evidence for Intelligent Design. In fact someone else, not Behe, has found a working bacterial flagellum that's simpler than the one Behe mentioned in *Darwin's Black Box*.

Phase II of the Wedge Document, though, has been very successful. Phase II reads in part:

> The primary purpose of Phase II is to prepare the popular reception of our ideas. The best and truest research can languish unread and unused unless it is properly publicized. For this reason we seek to cultivate and convince influential individuals in print and broadcast media, as well as think tank leaders, scientists and academics, congressional staff, talk show hosts, college and seminary presidents and faculty, future talent and potential academic allies.

Many of these goals have certainly been met. Congressional staff and talk show hosts have been especially successful. The Academy has been a bit tougher to crack. Dembski doesn't work at Baylor any more. Behe's department at Lehigh University issued a press release stating explicitly that they do not support his views on Intelligent Design. It's still on their website and it's a doozy. http://www.lehigh.edu/bio/News/evolution.html

This combination of scientific and scholarly expertise and media and political connections makes the Wedge unique, and also prevents it from being "merely academic." Other activities include production of a PBS documentary on intelligent design and its implications, and popular op-ed publishing.

> "Alongside a focus on influential opinion-makers, we also seek to build up a popular base of support among our natural constituency, namely, Christians. We will do this primarily through apologetics seminars. We intend these to encourage and equip believers with new scientific evidences that support the faith, as well as to "popularize" our ideas in the broader culture."

All of that, they have done with a vengeance.

Remember back when the Big Bang was a new idea and they had church seminars to teach people about red-shifted galaxies?

No? Me either.

It didn't happen.

Remember how they got senators and representatives on board to pass resolutions and laws requiring the teaching of the Big Bang in public schools? No?

Well, do you remember when school boards and science standards committees insisted on policies to teach scientific criticisms of Steady State theory?

It didn't happen.

Remember when a coalition of scientists hired a publicist and attempted to popularize their ideas in the broader culture? A publicist? Do theoretical physicists and cosmologists even know what a publicist is?

No, they didn't do any of that.

It wasn't on the list of goals and objectives. They took their measurements, wrote their papers, had their conferences and piled up the evidence. When a preponderance of scientists were convinced it was true, the Big Bang started showing up in high school text books without a constitutional crisis and without legislative pressure. The religious implication of the Big Bang was very clear, but even the dreaded ACLU wasn't interested.

What makes Intelligent Design so special?

The Big Bang proved itself as science. Intelligent Design has not. It's not science. It's religion. When I engage in on-line debates and my opponent talks about the designing intelligence, the guiding intelligence, the designer, or whatever, I refuse to go along with the euphemisms. I tell them "you are talking about God. You know it. I know it. Nobody is fooled."

Once in a while I get asked what scientists are so afraid of. Why do they so vehemently fight against Intelligent Design? What about it makes them so mad? Even William Dembski asks that question every now and then.

It's a very revealing question. The answer is supposed to be because all scientists are atheists and ID proves the existence of God. The question reveals that Intelligent Design is a religious argument, and has nothing to do with science.

I'll give you what, in my opinion, is the real reason Intelligent Design makes scientists, if not frightened, at least angry. Science must be free to go where the evidence takes it. If it is required to conform to religious correctness it will die. If biology must be religiously correct there will be no investigation into abiogenesis or the history of life. There will be certain scientific questions which scientists will be required to answer with "God did it" and then turn away. Almost any scientist alive would fight against that tooth and nail. It's not the God part that's offensive. Lots of scientists believe God is the author of the entire universe. It's the turning away that cannot be tolerated. It's the potential requirement that ignorance must be embraced and accepted that turns otherwise gentle zoologists and biochemists into the intellectual equivalent of wolverines.

Our wealth, our health and our culture rest on science. If all science—and make no mistake it will not stop with biology—is required to be religiously correct, science will be destroyed. Our culture, which rests on science, will also be destroyed.

We can't let that happen. Nature is beautiful. It's splendid. Science is a powerful tool which allows us to look into nature's secrets. To end that quest of discovery would be a monumental tragedy. There are just too many fascinating things out there to see and to learn.

We know now from the massive mountains of evidence that have been piled up over the last 150 years that Darwin was right when he said "from so simple a beginning endless forms most beautiful and most wonderful have been, and are being, evolved."

"Evolution: It's Only a Theory"
And Other Famous Creationist Arguments

This has always been one of my favorite subjects. I spent about fifteen years in a long conversation with creationists hashing out this topic. When I first had an internet connection, it didn't take me long to find usenet. Shortly after that I stumbled across Talk.origins. I quickly became fascinated with the history of life. I had always believed that evolution was true, and I'd always known there were creationists, but I didn't know any of the creationist arguments. After I began to debate them, I discovered to my horror that some of the things I thought I knew about evolution turned out to be creationist propaganda. I enjoy being hoodwinked about as much as the next person and so I launched into an extensive and pretty much ad hoc study of evolution.

I've learned a lot since those days and I'm delighted to still be learning. I have told many creationists that the evidence for evolution is so vast it cannot be examined in detail in a single human lifetime. It's a nice rhetorical flourish, if I do say so myself, but it also happens to be my personal experience. I'll never know it all and I'm glad I'll never come to the end of it.

What's really interesting about creationism—and I think you may see the pattern as I go through these creationist arguments—is that they know in their hearts that they will be required to destroy every single detail of science in order to refute evolution. This is especially ironic when done on the internet using a computer.

But let's get started. Many of these creationist arguments are cut and pasted directly from creationist websites. I wouldn't want you to think I'm misrepresenting them.

This is from a website called "The Truth Seeker" (http://www.thetruthseeker.co.uk/?p=204)

> "Darwin's theory of evolution is exactly that: a theory, with little in the way of hard scientific evidence to back it up. Yet it has been accepted almost as an article of faith in the modern world."

This is called the equivocation fallacy. You take a word that has more than one meaning and you use it one way while pretending to use it in another. Most people use the word "theory" to mean "speculation" or "guess." But scientists use the word "theory" in a very different way. Generally in science it is defined as "an organized system of accepted knowledge that applies in a variety of circumstances to explain a specific set of phenomena." Taken that way, of course, there's no speculation involved at all. So when the creationist textbook stickers in Georgia said "is a theory, not a fact, regarding the origin of living things" it is clear that the equivocation fallacy was being committed. (https://en.wikipedia.org/wiki/Selman_v._Cobb_County_School_District) It is my belief that it is used cynically, to imply that evolution is merely speculation and therefore false.

Obviously if The Theory of Evolution is merely speculation, then so is Atomic Theory, Germ Theory of Disease and the Theory of Gravity. I'm afraid that for most fundamentalists and creationists those things are mere speculations. Steven J. Gould addressed this issue in his essay "Evolution as Fact and Theory":

> "[E]volution is a theory. It is also a fact. And facts and theories are different things, not rungs in a hierarchy of increasing certainty. Facts are the world's data. Theories are structures of ideas that explain and interpret facts. Facts do not go away when scientists debate rival theories to explain them."

Evolution is a fact. The Theory of Evolution explains the fact of evolution.

Creationists say "How can you say evolution is a fact? It has never been observed. You weren't there. Only God was there." Creationist Ken Ham is very fond of this one.

> "So during your witnessing attempts," he says. "When someone tries to write off the Bible as just a collection of myths, remember that history agrees with Scripture. In fact, only God was there in the beginning. He ought to know what happened."

Using this logic we would have to free everyone in prison who was convicted solely based on forensic evidence. If there were no human witnesses, then all of the fingerprints, DNA, ballistics, and so on would have to be thrown out. Obviously that's silly. It's another nice rhetorical flourish.

It's also false.

The first instance of speciation was observed by Hugo de Vries in 1905. http://www.macroevolution.net/hugo-de-vries.html Since that time many instances of speciation have been observed in both the wild and in the lab. The instances of so-called "microevolution"—variations in a gene pool over time—have been observed a host of times. Historically, the fossil record is very clear that through time species have evolved. Though you can't go back and film the entire history of life, when you see a group of fossil hominid skulls arranged simply by age you can see a general progression from ancient forms that no longer exist to modern forms that we see around us today. The same is true of horse fossils, whale fossils, fish fossils and so on.

This leads to another common creationist argument: there are no transitional fossils. Ken Ham, while attempting to refute an article on transitional fossils by Boyce Rensberger http://www.stephenjaygould.org/ctrl/news/file021.html said

> "Rensberger avoids discussing the vast gulf between non-living matter and the first living cell, single-celled and multicelled creatures, and invertebrates and vertebrates. The gaps between these groups should be enough to show that molecules-to-man evolution is without foundation."

At talk.origins there is a running joke that gets brought up every time a new fossil is found. "Oh no!" they say in mock horror, "Two more gaps in the fossil record!"

Creationists also like to say that a transitional fossil would be a half-dog, half-chicken affair that couldn't exist and wouldn't have lived very long if it did. Every fossil, they say, is a complete animal and therefore not between any two other animals. Defined that way, of course there are no transitionals. But you don't get to make up your own private definitions of words. There are many examples of fossils that are a mosaic of features from older and more modern species. There are many finely graded transitional fossil series. There is one series of diatoms in which you can actually see the point where the new species developed.

When creationists say that evolution has never been observed they generally will accept the fact of insects evolving to cope with pesticides or bacteria evolving resistance to antibiotics. What they object to is what they call "macroevolution" or as Ken Ham likes to say it: "molecules to man."

My flippant answer to this is if my car can go one mile, it doesn't take a special magic for it to go one hundred miles. They never buy it. I usually have to find pictures of transitional fossils and begin to talk about the age of the earth.

Remember when I said creationists have to refute all of science? When it comes to the age of the earth, they have to claim that all physicists—not just the evolutionary biologists—are atheist liars. They do things like point to creationist "researchers" who have carbon dated live animals and fresh lava beds and shown that the tests concluded they were very ancient.

I have to patiently point out that carbon dating only works on artifacts younger than 50,000 years old. That you can get anomalous dates on all kinds of things but that doesn't discredit the entire method. I point out that Biblical archaeologists use carbon dating all the time and everyone believes their results. They usually aren't trying to talk about carbon dating though.

Actually, they are confusing carbon dating with radiometric dating—some creationists call it "radio dating." Since I barely understand the physics better than they do, I point out that the same physicists that give us microwaves and computers know what they are doing when they date the earth. It's a crude argument, but a more sophisticated one is rarely needed. Fortunately the people who do understand the physics are arguing on my side!

Charles Darwin wrote in the *Origin of Species*:

> "To suppose that the eye with all its inimitable contrivances for adjusting the focus to different distances, for admitting different amounts of light, and for the correction of spherical and chromatic aberration, could have been formed by natural selection, seems, I freely confess, absurd in the highest degree."

I bet I've argued the evolution of the eye hundreds of times. It's a creationist favorite. It's called the argument from extreme perfection. Darwin was aware of it—it's been around for a long time—and devoted an entire chapter in *Origin* to it. This argument has found new life with Michael

Behe and the Intelligent Design movement. Of course the argument boils down to "I don't understand it, so it can't be true!" Creationists don't seem to ever see the arrogance in that argument even though I have taken pains to point it out to them.

The sentence from *Origin* that I quoted above is an introductory sentence. That sentence is repeated in hundreds, maybe thousands, of creationist websites and books. They never quote the rest of the paragraph, though. Here it is:

> "Yet reason tells me, that if numerous gradations from a perfect and complex eye to one very imperfect and simple, each grade being useful to its possessor, can be shown to exist; if further, the eye does vary ever so slightly, and the variations be inherited, which is certainly the case; and if any variation or modification in the organ be ever useful to an animal under changing conditions of life, then the difficulty of believing that a perfect and complex eye could be formed by natural selection, though insuperable by our imagination, can hardly be considered real. How a nerve comes to be sensitive to light, hardly concerns us more than how life itself first originated; but I may remark that several facts make me suspect that any sensitive nerve may be rendered sensitive to light, and likewise to those coarser vibrations of the air which produce sound."

Since Darwin's time those numerous gradations have been found in animals all around the world, every step of eye evolution from a mass of light sensitive cells, to complex eyes even better designed than our own. Each grade is useful to its possessor. Clearly variation and natural selection can make an eye. It's not absurd in any degree.

I'd like to talk a bit more about that introductory sentence of Darwin's above. Imagine you are a creationist surfing the web looking for ammo with which to defeat the forces of Satan. You come across that Darwin quote—it's not hard to do.

You think to yourself "Darwin doubted his own theory! Who knew!" Then you read a little further and you come across this quote from Stephen J. Gould:

> "The extreme rarity of transitional forms in the fossil record persists as the trade secret of paleontology..."

There you have it from one of the arch evolutionists of all time! Evolution isn't true. And look at this:

> "No wonder paleontologists shied away from evolution for so long. It seems never to happen."

Niles Eldridge said that! How is that possible? Ernst Mayr said this:

> "Following phyletic lines through time seemed to reveal only minimal gradual changes but no clear evidence for any change of a species into a different genus or for the gradual origin of an evolutionary novelty. Anything truly novel always seemed to appear quite abruptly in the fossil record."

Can you believe it? The famous evolutionary biologist Ernst Mayr was secretly a creationist! How can this be?

I can now laugh about this little creationist trick, but it made me really mad when I first encountered it. These quotes are out of context. The creationist will give you the first sentence of the paragraph and not the rest of it. Or they'll quote the first part of a paragraph and, without any indication that there is excised material, continue the quote from paragraphs or pages later. I have met many creationists who refuse to use this technique. They know it's lying and to their credit it makes them uncomfortable. But almost every book and website attempting to refute evolution will use these out-of-context quotes as if citing the authorities just as anyone does in a scholarly paper. They don't care if you look the quotes up. If you are willing to do that, you are not who they are aiming at anyway.

Creationists use several types of arguments that are metaphysical and completely outside of science. They really can't see a difference. I will cover only two of them here.

One of the arguments is that evolution is random. "I don't want to believe that I'm an accident," they will say. They really don't like the idea of life simply rolling along by itself. God needs to be in charge and evolution says that he isn't.

There is a little book that's called *Fatal Flaws: What Evolutionists Don't Want You to Know* by Hank Hanegraff. It's the usual collection of creationist stuff, some of which I have presented here for you. In the entire book, natural selection is mentioned only once or twice in passing and not at all in the chapter entitled "Chance." Hannegraff and other creationists generally avoid the subject of natural selection. I think it's for at least two reasons. One is that it makes perfect sense and you can observe it going on all around you. The second is that it removes the element of chance from evolution.

It's true that we don't know what causes mutations and they appear by chance as far as we know, but natural selection, in the context in which it is taking place, is very far from random.

Creationists intuitively know that. They know it would be tough to refute and therefore they generally prefer to pretend it doesn't exist. They generally focus instead on simply declaring evolution is random and then working on the odds of a cell, or whatever, appearing randomly by chance. I don't understand the math involved, but I don't need to. Those calculations are based on the false assumption that natural selection is random. The entire argument about the odds of this or that evolving is simply a red herring.

Theistic evolutionists believe God guides the process of variation and natural selection, but Biblical literalists don't like that argument. If you allow that, then you have to allow all of evolution. Hanegraff contends that theistic evolutionists are an oxymoron like the expression "flaming snowflakes" and, like most fundamentalist creationists, merely pretends the theistic evolutionary argument doesn't exist.

The other metaphysical argument is that evolution is an argument for atheism.

Here is creationist Philip Johnson beating one of his favorite straw men http://www.wsj.com/articles/SB934759227734378961

> "The reason the theory of evolution is so controversial is that it is the main scientific prop for scientific naturalism. Students first learn that "evolution is a fact," and then they gradually learn more and more about what that "fact" means. It means that all living things are the product of mindless material forces such as chemical laws, natural selection, and random variation. So God is totally out of the picture, and humans (like everything else) are the accidental product of a purposeless universe.

Most of the "evolution is atheistic" arguments are one step removed from the actual claim. That evolution is atheistic is merely assumed and then all further argument is based on that and become arguments against atheism. Atheism is an opinion, not science and I won't debate it.

There is one related argument that I do debate and it goes like this: If evolution is true and we just evolved from animals there's no reason for morality let alone God."

This argument warms the cockles of my heart. Because the main thrust of this argument is that morals have no value in and of themselves. They are something imposed on us by God just like your mother making you wear a sweater when she is cold. Without a god making us be moral, there's no point to morality. Obviously I don't think so and I quickly point out the flaw in their reasoning. They usually backtrack very quickly or refuse to discuss the topic.

If we evolved, we evolved as a social species and our morals evolved right along with our bodies. Even if that were not true, morals clearly have an intrinsic value. Behavior has consequences and good behavior has good consequences. Being honest, truthful and kind makes life nicer, not worse. The most hardened creationist has no trouble grasping that.

There are a lot of creationist arguments that I'm not going to cover in this essay—"evolution is a religion, not science"; "evolution violates the 2nd law of thermodynamics"; "archaeopteryx is a fraud"; "human and dinosaur footprints are found together"; "Niagara falls proves the earth is young"; "moon dust proves the earth is young" and on and on. It's been a long time since I've seen a new creationist argument. I keep hoping they will evolve. No such luck!

I'll let St. Augustine of Hippo have the last word here:

> "Usually, even a non-Christian knows something about the earth, . . . about the kinds of animals, shrubs, stones, and so forth . . . Now, it is a disgraceful and dangerous thing for an infidel to hear a Christian, presumably giving the meaning of Holy Scripture, talking nonsense on these topics; and we should take all means to prevent such an embarrassing situation, in which people show up vast ignorance in a Christian and laugh it to scorn. . . . If they find a Christian mistaken in a field which they themselves know well and hear him maintaining his foolish opinions about our books, how are they going to believe those books in matters concerning the resurrection of the dead, the hope of eternal life, and the kingdom of heaven, when they think their pages are full of falsehoods. . . ? Reckless and incompetent expounders of Holy Scripture bring untold trouble and sorrow on their wiser brethren when they are caught in one of their mischievous false opinions and are taken to task by those who are not bound by the authority of our sacred books."

14. References & Sources

Additional Websites

The Talk.Origins Archives
http://www.talkorigins.org/
This a large, readable and well organized website that is excellent at answering questions raised by creationists.

National Center for Science Education (NCSE)
http://www.ncseweb.org
Has many links and up-to-date resources and news items on evolution

Kenneth Miller's Website
http://www.millerandlevine.com/km/evol/

Understanding Evolution
Sponsored by the National Center for Science Education and the University of California Museum of Paleontology
http://evolution.berkeley.edu

Books

Origin of Species by Charles Darwin
Finding Darwin's God by Kenneth Miller
Darwin's Ghost by Steve Jones
The Wisdom of the Bones: In Search of Human Origins by Alan Walker, Pat Shipman (contributor)
Lucy: The Beginnings of Humankind by Donald Johanson
The Sacred Depths of Nature, by Ursula Goodenough
What Evolution Is by Ernst Myer
The Selfish Gene by Richard Dawkins
The Blind Watchmaker by Richard Dawkins
The Beak of the Finch, by Jonathan Weiner
Biology 4th ed. by Solomon, Berg, Martin, Villee

Notes

www.ingramcontent.com/pod-product-compliance
Lightning Source LLC
Chambersburg PA
CBHW070326190526
45169CB00005B/1764